HEMP
BOUND

ALSO BY DOUG FINE

Not Really an Alaskan Mountain Man
Farewell, My Subaru: An Epic Adventure in Local Living
Too High to Fail: Cannabis and the New Green Economic Revolution

HEMP
BOUND

DISPATCHES FROM THE FRONT LINES OF THE NEXT AGRICULTURAL REVOLUTION

Doug Fine

CHELSEA GREEN PUBLISHING
WHITE RIVER JUNCTION, VERMONT

Cover painting, *Hemp Fields*, by Richard Fields.
To order prints of *Hemp Fields*, contact morse@hawaii.rr.com

Project Manager: Bill Bokermann
Project and Developmental Editor: Brianne Goodspeed
Copy Editor: Laura Jorstad
Proofreader: Helen Walden
Indexer: Margaret Holloway
Designer: Melissa Jacobson

Printed in the United States of America.
First printing March, 2014.
10 9 8 7 6 5 4 3 2 1   14 15 16 17

Our Commitment to Green Publishing

Chelsea Green sees publishing as a tool for cultural change and ecological stewardship. We strive to align our
book manufacturing practices with our editorial mission and to reduce the impact of our business enterprise in
the environment. We print our books and catalogs on chlorine-free recycled paper, using vegetable-based inks
whenever possible. This book may cost slightly more because it was printed on paper that contains recycled
fiber, and we hope you'll agree that it's worth it. Chelsea Green is a member of the Green Press Initiative
(www.greenpressinitiative.org), a nonprofit coalition of publishers, manufacturers, and authors working to protect the world's endangered forests and conserve natural resources. *Hemp Bound* was printed on FSC®-certified
paper supplied by Thomson-Shore that contains 100% postconsumer recycled fiber.

Library of Congress Cataloging-in-Publication Data
Fine, Doug.
  Hemp bound : dispatches from the front lines of the next agricultural revolution / Doug Fine.
     pages cm
  Includes bibliographical references and index.
  ISBN 978-1-60358-543-9 (pbk.) — ISBN 978-1-60358-544-6 (ebook)
  1. Hemp. 2. Hemp industry. I. Title.
  SB255.F56 2014
  633.5'3—dc23
                                2013048926

Chelsea Green Publishing
85 North Main Street, Suite 120
White River Junction, VT 05001
(802) 295-6300
www.chelseagreen.com

*F*or Nancy, Dee, and Kate, who, in one of the hot springs that increasingly seem to provide my office, made sure I asked about R-value (remember this, IRS, if any of this year's deductions seem esoteric). And for Bill Althouse, for the carbon-neutral limo ride.

Contents

Hot Off the Hemp Presses— Just Trying to Keep Pace with America's Light-Speed Return to Its Usual Attitude About a Plant

*H*emp cultivation is about to become legal (and shortly thereafter, big) again in the United States. It started to happen while I was about halfway done with this book. I'm just not used to winning big, important societal battles outright. It's an astonishing no-brainer. And it directly affects my life.

To give just one example, my plan the day hemp becomes legal is to begin cultivating ten acres of the plant so that my Sweetheart no longer has to import from China the material she already uses to make the shirts I wear in media interviews to discuss the fairly massive economic value of hemp. In a cynical age, we can use one less irony.

It's not an exaggeration to say that in humanity's eight-thousand-year relationship with the hemp plant, this past year has been the most impactful one since the first Paleolithic hunter with blistered feet noticed that hemp's fibers made a stronger sandal than the leading brand. We saw Kentucky's passage of hemp (also called industrial cannabis) legislation and the Colorado legislature's near-unanimous approval of commercial hemp cultivation in time for the 2014 planting season[1] (making ten states that have in some form allowed cultivation, including North Dakota and Vermont).

Most important, the U.S. Congress, as I send this book to my publisher, is also poised to re-legalize domestic hemp cultivation without the necessity of federal approval for the first time since 1937.[2] The federal drug war is the last impediment to U.S. farmers and entrepreneurs benefiting from what we'll see is already a half-billion-dollar hemp industry in Canada.

In the House version of the massive 2013 FARRM Bill (one of those "must pass" infrastructure bills that come along every five years), the psychoactively inert hemp plant was removed from the purview of the federal Controlled Substances Act (where it currently is classified as more dangerous than meth and cocaine). The wording of a bipartisan amendment allows industrial cannabis cultivation as part of university research in states that permit hemp farming.

To many who have battled for years to get the plant we're going to be discussing back into the economy, this seems like a baby step. But Canada followed a similar course prior to ramping up its hemp crop in 1998, conducting government-sponsored research into the best cultivars (seed varieties) for its farmers to use from Ontario to Alberta, starting in 1994. In Europe as well, new cultivar certification is a three-year process.

"We don't have the seed stock in the U.S. anymore," plant breeder, former Big Corn scientist (he invented the plant molecular marker), and renowned hemp researcher David West told me. "I checked with the National Seed Storage Laboratory. They found me a few bags of the old Kentucky seed sitting in a hallway. They were long rotten. It's hard to express what a terrible loss that is—this was a blending of Asian and European cultivars that comprised the best hemp germplasm in the world."

In other words, it'll take some time and study to re-learn what hemp varieties will work in the American heartland's soil. West and many others believe that feral hemp—the "ditch weed" that's survived prohibition in places like Nebraska—might be a source for rebuilding the stock for a seed that's been cultivated in the New World since the earliest colonial moments (1545 in Chile, 1606 in present-day Nova Scotia[3]). Let Darwin pick the cultivar, is the argument. "I tell farmers to make clandestine collection of feral seeds in advance of legalization," West told me with a laugh.

What's astonishing, for someone who's been following "drug" policy pretty much full-time for three years now, is that it's really happening: America's worst policy since segregation—cannabis prohibition—is ending. It lasted seventy-seven years. That's how powerful the rhetoric of the drug war has proven since alcohol prohibition's chief crusader, Harry Anslinger, took the helm at the new Bureau of Narcotics in 1930. Using the same arguments that the American people no longer bought regarding beer, cannabis prohibition began seven years later. Banning (in actuality, restricting to zero by not issuing federal permits) hemp cultivation is like banning wheat. It's a surreal policy. Absolutely, 100 percent baseless.

As recently as 2012, hemp-friendly legislation would get floated by the leading congressional iconoclast every couple of years, and promptly get laughed out of committee. For three-quarters of a century, for my father's entire lifetime, the truth didn't matter, until last June. Hemp burst back from within our genetic memory. I'm

HEMP PIONEERS

Dr. David West, Geneticist, Actual Twenty-First-Century American Hemp Researcher

From his home alone in Prescott, Wisconsin, along the St. Croix River, the sixty-five-year-old West said that what is, on the surface, his unusual journey from legendary Big Ag researcher to legendary hemp researcher actually follows what for him was an obvious course. After pioneering the use of molecular markers in plant breeding (now part of the standard commercial plant identification tool kit), he "watched the seed industry get taken over by the chemical industry" in the 1990s. At the same time, he told me, "One day I saw a helicopter land in a neighboring field [in Wisconsin] to eradicate feral hemp. Now, as a plant breeder, I'm quite aware of what hemp is. I thought, *What the hell is going on here?*"

In a 1994 paper titled *Fiber War*, West declared modern hemp's agricultural value, a radical view for someone with unimpeachable creds in what was fast becoming the GMO monoculture world (which industry he had by this time left, saying, "I don't want to grow terminator corn for Monsanto").

His notoriety from that piece and subsequent writing on hemp (as well as the co-founding of the North American Industrial Hemp Council) led, in 1999, to him being contacted by Hawaiian representative Cynthia Thielen (R-Oahu), who is still in office and to this day battling for hemp production in the Aloha State. She's trying to find a replacement for the declining sugarcane industry, help remediate soil, and find an animal feed that can be grown on the islands.

"Cynthia basically said, 'Do you want to come to Oahu to grow hemp?'" West explained. It was an *Is this a trick question?* moment. Or as West, who thought that maybe he was on *Candid Camera*, puts it, "It was a once-in-a-lifetime opportunity."

The half-acre project, which got its federal permits to acquire hemp seed at what West described as "the last second," ran from 1999 through 2003, and began with a state Hemp Day declared by the governor for the morning of the first planting on December 19, 1999.

"There were cameras, the Kahuna ceremonial blessing, the whole deal, then everyone went home, and it was me on a fenced-in, alarmed patch of dirt, dealing with every problem farmers have always had to deal with."

One of his first discoveries, he told me with an *It's funnier now* snicker, was that "birds love hemp. Took a while to rig a netting system that kept them out. They ate the whole first planting."

The project was funded by a hair care company interested in a publicity stunt for what West called a "dash" of hemp oil in its product. West was fine with that. "When I saw like-minded people at energy fairs speak about hemp without any real knowledge—and how could they have knowledge?—I realized that what we really needed was some studies. But I also knew, since I worked for seed companies for years, how much that kind of research costs."

West groaned like a hungry man when I told him I was just back from a visit to the sixty acres of hemp that Colorado farmer Ryan Loflin was able to cultivate in 2013. "My study was on a very small, academic scale, but we wound up showing that hemp could be viable in Hawaii's latitude, which is important because growing hemp is all about the photoperiod. It was a Chinese cultivar that worked best. Grew more than ten feet tall. And it was on former Dole plantation land."

When the project wrapped, West said, George W. Bush was in the White House, 9/11 had happened, and almost no one was paying attention, not even the reps at Alterna, the hair care company. But he rigorously recorded his data and methods, because "I knew people would care about hemp again."

West calls himself retired today, but he's still researching ditch weed in Nebraska and trying to fund a hemp genome project. I watched a YouTube video where he visited a feral Midwest hemp field, dissecting several plants' morphology. His explanation for all the activity is that it's involuntary. "When you open one door with hemp genetics and even with American hemp history, a dozen other doors open," he said.

He then launched into perhaps a twelve-minute (and riveting) tale of one of the original "hempreneurs," David Myerle, who went bankrupt in the 1820s after planting and contracting for hemp in Kentucky, Missouri, and elsewhere.[4] One problem? The U.S. Navy kept rejecting his multi-ton hemp deliveries, either for quality reasons or because of corrupt ties to another supplier. A bigger problem? Too many of his workers were dying of pneumonia trying to implement his special water-based hemp-processing regime. We'll be looking at less deadly and potentially faster methods in these pages.

glad my pop's getting to see the dawn of the era in which America returns to a healthy agricultural paradigm. We all have a significant vested interest in its success.

Anthropological and archaeological research, including the recent discovery of cannabis in a twenty-seven-hundred-year-old Chinese tomb, shows that we humans probably have been making use of this plant at least as long as we have any other. It's the last seventy-seven years that have been atypical. In the Shinto coronation ceremony, for example, Japan's incoming emperor wears a hemp robe, and not to prove a political point, but because of the plant's broad value. It symbolizes abundance, comfort, and health.

Places like China, Romania, and France never stopped cultivating hemp. And as I and others have pointed out, even U.S. industrial cannabis prohibition got off to a poor start soon after the federal drug war got rolling with the passage of the Marihuana Tax Act of 1937.

What happened was, in 1942, the U.S. Navy tried to place an order for the (far and away) best rope, on which it and America's war effort were dependent—they needed as much as twenty tons of rigging per vessel. For some reason there was none in the supply room. Seems Japan had captured the new Filipino source—notice that the drug war had already shifted American business offshore.

So the United States Department of Agriculture (USDA) quickly shot and released a passionately pro-hemp documentary, a classic of short-form propaganda vérité. The film begs farmers to plant as much hemp as possible, yesterday. It's called *Hemp for Victory*. Check it out on YouTube.

An analysis of the reasons behind today's unfolding sea change in U.S. public opinion about drug policy is a book in itself, but suffice to say that this is a people-driven hemp economic boom we're about to experience (actually we're already experiencing it, but Canada's making most of the dough). And it's about time.

Not only do we Americans buy that half billion dollars of Canadian hemp products every year, but the number is growing 20 percent annually. We're just not allowed to grow it here. "This kind of trade imbalance is why the American colonies fought for independence from Britain," Colorado rancher and putative hemp farmer Michael Bowman told me just as he was about to violate federal law and throw a few seeds on the ground on July 4, 2013. The date not being accidentally chosen.

The Drug Enforcement Administration (DEA), whose two-and-a-half-billion-dollar budget you and I pay, enforces the Controlled Substances Act. Cannabis is the largest target of the drug war, by far. The agency's deciders are putting its budget ahead of the clear interests of the nation, fighting tooth and nail to defend the outrageous hemp ban.

I should note here that, after three years of reporting from the drug war's front lines, I believe the good men and women of the DEA and other law enforcement agencies are doing their best. I applaud efforts to stem the flow of dangerous drugs like cocaine and black-market prescription pills. What's coming through here in the hemp discussion is my citizen frustration with a government agency that, for the good of the country, needs to make an immediate 180-degree shift on hemp. As we'll see, the agency can become part of the industry's regulatory process. That's how Canada does it.

Instead, as usual, the DEA's lobbyists brought all the now conventional lies to the 2013 congressional hemp legalization discussion—*People can't tell the difference between hemp and psychoactive cannabis*; *People might smoke their drapes*—only this time it didn't work. When Representative Jared Polis (D-Colorado), along with his bipartisan friends Thomas Massie (R-Kentucky) and Earl Blumenauer (D-Oregon), brought forth their FARRM Bill amendment on July 11, 2013, it passed by a vote of 216–208, with 69 Republicans voting yea.

On the Senate side, there was also some delicious Beltway cloakroom strong-arming surrounding the garnering of Senate Minority Leader Mitch McConnell (R-Kentucky)'s support for hemp, best reported by Ryan Grim in the *Huffington Post*. Allegedly, McConnell's pro-hemp Bluegrass State colleague, Rand Paul, promised not to back a McConnell Tea Party primary opponent if the senior senator threw his support behind industrial cannabis.

Long story short, according to Eric Steenstra, president of the Vote Hemp advocacy group, after thirteen years of hemp lobbying, he's suddenly noticed that "hemp hasn't been controversial" in recent legislative discussion.

So even though Senate and House FARRM Bill negotiations timed out during the 2013 federal government shutdown, I think hemp legalization will have happened, if not by the time you're reading these words, then by mid-2014.[5] Sure is proving to be a page-turner, though, complete with six-month-long cliffhanger.

From the perspective of a patriotic American who's just researched hemp's potential from Canada to Hawaii, Germany to Colorado, things are moving from fantasy to reality so quickly that it's kind of making me believe in a societal version of *The Secret*—ask for what you think's best for your nation's economy, the planet at large, and your children's future, and you will get it.

My excitement is perhaps best described this way: Ten months ago, I was shocked that Congress was even discussing hemp seriously. Suddenly I'm confident that future editions of this book will be printed on U.S.-grown hemp paper.[6] In fact, in these pages we'll be meeting two of the farmers who will be making it happen. It's been a dream since my I wrote my first book that not a tree would have to perish in order for me to publish—particularly since the idea of a sustainability author printing on shredded forests for some reason felt a little ticklish to me. Smarty-pants audience members were always asking me about that at live events, especially at those dang college talks. Now it looks like the paper itself will soon be soil fixing.

Humans, after a seventy-seven-year break, are returning to one of the most useful plants ever bestowed on them. And it happened while I was in the middle of writing about said plant, so I had to stick this note in here to hammer home the point that by the time this book hits shelves and e-readers, we might have hemp drapes in the White House Situation Room.[7]

I mean for practical reasons. Hemp fabric is less flammable and longer lasting at a lower cost than the leading brand. So when you see farmers, energy companies, and policy makers from places like North Dakota and Kentucky expressing outrage in these pages about their inability to capitalize on the production side of the exploding worldwide hemp phenomenon, you can bet they're rubbing their palms together now, just a few months later.

That's because the U.S. market's well ahead of the politics. It is expensive to have to import hemp. The plant is popular enough to do it, but it'll be a pleasure not to have to, folks in the business tell me. Which is to say, people are already making real wampum from hemp.

As John Roulac, founder and CEO of Richmond, California–based Nutiva, the seventy-million-dollar company that makes omega-balanced and mineral-rich hemp seed oil, puts it, "Our company has doubled in size each of the past two years, has been growing 41 percent per year since 2006. *Inc.* magazine named us one of its fasting-growing companies in 2010. That's only going to continue. Look for hemp to grow fencerow-to-fencerow in the heartland. It's going to displace the corn and soy duopoly in the American Midwest."

While I've got you thinking big picture, and before we launch into the plant's most lucrative digital age killer apps, I thought it might be helpful to include a quick, explicit definition of hemp. That's because it's finally sunk in, after several years spent researching an industry that's indeed growing 20 percent a year (and that's just the hemp seed oil market), that such growth means a lot of new people are coming to the topic all the time. These folks will want to know what exactly this plant is we're discussing.

Even some of 2014's U.S. farmers will be fresh to the species, given that Canada's cultivators can't keep up with demand, meaning a much-needed cash crop is ready to roll Stateside. Now. The North American industry is growing like feral hemp in a Nebraska ditch.

Hemp[8] includes all varieties of the *Cannabis* genus that contain negligible amounts of THC (a component of the cannabis plant that can be intoxicating when heated). It is synonymous, as we've said, with industrial cannabis and in fact has been used in industry for so long that linguists can trace when and where key language changes occurred based on a culture's word for the plant.

Cannabis is a dioecious plant (there are males and females), and branches are covered with hand-like leaf fans. Originating in Central

HEMP PIONEERS

John Roulac, Founder and CEO, Nutiva

Given that I've been pouring a tablespoon of Nutiva's organic hemp oil (Canadian-grown, for now) into my family's breakfast shake every day for half a decade (to the tune of about eight hundred dollars per year and willingly counting), I thought it worthwhile to ask the company's fifty-four-year-old founder about his personal and entrepreneurial journey. Turns out his arc is similar to that of a solar electrician friend of mine in New Mexico, who's so busy that he describes himself as a "failed hippie."

"I was a forest activist in the California redwoods in the 1980s and early '90s," Roulac told me. "And the opponents would say, 'If you're not gonna cut down trees, where will our houses come from?' That led me to hemp fiber, one of the strongest in the world. Then I discovered that the seed is one of the most nutritious available."

That discovery still moves Roulac profoundly, judging by the fact that for about the next eight minutes I couldn't type fast enough to keep up with the guy's love song to hemp oil. It's making him rich—the WE'RE HIRING button on the privately held company's home page is large—but clearly Roulac was feeling it.

Highlights from his serenade include this, when I asked how hemp oil compares with other omega-rich oils like flax: "Flax is fine, hemp oil is divine. Hemp has what flax, chia, and fish oil don't: both GLA [gamma linolenic acid] and CLA [conjugated linoleic acid]—omega-6 fatty acids that are superfoods. GLA is an anti-inflammatory, and CLA is a building block of cell membranes, to just scratch the surface on those two. So hemp has a better fatty acid profile than flax. The shelled hemp seed—the hemp heart—is a gift from the universe. One little seed gives you magnesium—a master mineral involved in three hundred chemical processes in the body—zinc and iron. Vegans in particular can be short on those. Hemp is just nutritionally superior to flax and will surpass flax sales in the coming decade."

And it went on like this for a while. Let me tell you, as someone who finds living preferable to the alternatives, I was all ears.

The business side kind of blends with the societal side with Roulac, and on both counts you can't accuse the fellow of failing to think big. "Our goal is to change the way the world eats, and to improve the food systems across the food chain. And we're already doing this."

How so? "Today we're working with states like Kentucky to get hemp grown domestically. I testified there," he told me. "But our biggest issue is that we only sell certified organic seed and oil, and there isn't the infrastructure yet with hemp. Believe it or not, even though GMOs are banned in Canadian hemp, which is a nice gesture, today most Canadian farmers are GMO farmers who use hemp as a bridge crop for three months and there's plenty of pesticides applied the rest of the year at least. It's part of the GMO cycle. We working to build that organic market."

For that reason, toward the end of our conversation, Roulac added a challenge to consumers: "If you want to see a green future, buy organic hemp. The more organic hemp you eat, the more organic hemp will be planted, and the healthier the planet will be."

Now, I'm a journalist of some experience, and I recognize a line out of an industry trade group playbook when I hear it. But I'll cut Roulac some slack and include his talking point, for two reasons. First off, he's talking about a plant that it is at the time of this writing a federal felony to cultivate. And second, as a sustainability writer for two decades who's just back from visiting a lot of hemp farms and reading a lot of hemp research, and as a fellow who's had to work with drought-affected soil on my own ranch, I can tell you he's right.

Asia, hemp has a four- to six-month growing cycle and has been successfully cultivated on every continent except Antarctica. Mature plants range from two to twelve feet tall, depending on variety.

There are two main reasons the plant is important to humanity: It feeds us and it protects us. The seed oil, as we'll see in a Canadian study I visited in the lab, is an incredibly nutritious protein- and mineral-rich source of essential fatty acids. And the fiber is freakishly strong. When you slice a hemp stalk in half, you'll see, nestled in a snug hollow tube, a long, string-like band running the length inside. This is hemp's famous bast fiber. Cultivated correctly, it's stronger than steel. So when I say hemp protects us, I mean it has done so from the time of our earliest and still most durable clothing and shelter right up to our next-generation body armor.[9]

Okay. So now as I rattle off all the hempsters' favorite historical uses for hemp (which, really, is a way of showing how patriotic and conservative their view is, how history is on hemp's side), you'll know the reasons why.

Thomas Jefferson famously wrote his Declaration of Independence draft on hemp paper, and Betsy Ross's first flag was made from the plant. Early Virginia colonists were ordered to cultivate industrial cannabis and could even pay their taxes with harvest shares. More recently, a 1938 *Popular Mechanics* article that's become legendary among hemp boosters presciently called hemp "the billion dollar crop," and extolled its bafflingly strong fibers' twenty-five thousand industrial uses, half a dozen of which we'll be looking at in this book.

Really all of this is most elegantly expressed in that Shinto coronation ceremony. The message is, "Bring the hemp with you." It's what anthropologists call a camp follower—it was toted nearly everywhere as humans traipsed around the globe.[10]

Which is actually good for the U.S. industry, according to hemp agronomist Anndrea Hermann. Rather than relying on feral Nebraskan ditch weed, "Farmers in Kentucky and Colorado can look for varieties that have worked in climates similar to theirs." An Uzbek cultivar, for example, might be perfect for Illinois, where as a state senator Barack Obama voted twice for hemp cultivation.

West told me that a century before *Hemp for Victory*, the U.S. Navy was so desperate for hemp rope that the federal government began contests, in the 1840s, for the production of high-quality fiber strains that could compete with the then-standard-bearing Italian and Russian varieties that taxpayers were being forced to expensively procure.

"Missionaries sent back Chinese hemp, farmers blended it with the more coarse European strains we already had, and the result was the finest hemp in the world," he said. "It's generally called Kentucky hemp, but there were many named varieties with specific properties that were well known and widely marketed for more than half a century. Government-run breeding programs continued until the 1930s." USDA-researched strains had names like Kymington, Keijo, and Yarrow.

And so was birthed an industry that employed thousands, earned millions, and spanned a dozen states by the turn of the twentieth century. Kentucky's first millionaire, Lexington's John Wesley Hunt, made his pile in the 1840s weaving his hemp crop into rope, according to the *Lexington Herald-Leader*.[11]

For a solar-powered goat herder like me, it's worth noting that, according to the Canadian government, hemp can be cultivated with almost no pesticides (though as we'll see, there's a bit more to that story than that).

I think that covers everything a newcomer should know before diving into this book. Oh, except for a special note to those very new to the topic—especially those who were raised, as I was, on "Just Say No" era rhetoric: You can't possibly confuse industrial cannabis with psychoactive cannabis, for a number of reasons.

For one, hemp grows in vast, dense fields of thin, stick-like plants (as opposed to the flower-heavy, manicured prima donna social/medicinal cannabis plants). Even more crucially, hemp's pollen will immediately ruin a psychoactive cannabis crop, by diluting the psychoactivity that, as President Obama so eloquently pointed out in describing his own affection for the plant, is "the point."

In fact, this is why when California passed a terrific bit of industrial cannabis legislation in 2011 with bipartisan and local law enforcement support (inexplicably vetoed by Governor Jerry Brown[12]), it called for cultivation only in counties (incidentally, very politically conservative counties) that are far from the Emerald Triangle region in the northern redwoods that's known worldwide for its top-shelf psychoactive cannabis.

Also, eating, drinking, wearing, or sleeping in a house made of hemp will never cause intoxication, nor THC to show up in a urine test. Young children can safely eat and drink hemp oil and other hemp food products. All are as healthy and inert as flaxseed and cod liver oils.

In the real world, would you care to know how many cases of hidden or accidentally psychoactive cannabis plants have turned up in the course of Canada's decade-and-a-half-old and burgeoning hemp industry, according to Hermann, who does the testing and inspecting in the province of Manitoba? Zero. "There's no confusion," she told me. "We've been at this for fifteen years now. Everyone recognizes hemp's great value to farmers and the country."

—DOUG FINE

Funky Butte Ranch, New Mexico

August 22, 2013

After-School Snacks Before Doritos

*T*urns out your Deadhead roommate was right. Sort of. It isn't so much that hemp, useful as we're about to see it is, will automatically save humanity. It's that the worldwide industrial cannabis industry can play a major role in our species' long-shot sustainable resource search and climate stabilization project. For that to happen, the plant must be exploited domestically in ways upon which the marketplace smiles. No pressure: We fail? We just go extinct. The Earth'll be fine.

Hemp hands us a ninth-inning comeback opportunity. At the same time that it stimulates community-based economic growth on the producer side (and not a little bit, if a farm community is serious about implementing some of the ideas we're about to discuss), large scale re-adaptation of one of humanity's longest-utilized plants will provide sustainable energy, regionally produced food, and digital age industrial materials on the consumer end.

The planet's struggling soil wins, too: All those farmers put back to work growing a viable cash crop? They're remediating soil toxified or desertified after a century of monoculture (or in Kentucky's case, coal mining).

"Since hemp is so good for soil structure," British hemp expert John Hobson told me of European use of the plant today, "it's utilized as a true [rotational] crop. Even when growing cereals is more profitable, hemp keeps those yields from going down and blots out weed invasion." Hobson should know: He advises farmers on which hemp cultivars his Lime Technology company wants grown for use as building material in the Continental construction industry.

The key to success, from humanity's perspective and from an economic perspective, is multiple use of the plant. This starts on the farm with a concept we'll be discussing called dual cropping, but which really should be called tri-cropping. Basically, one hemp harvest can and should be used at once for food, energy, and industrial components (like car parts, building insulation, and clothing). Hemp is already in BMW and Dodge door panels.

The fact is, after my most recent intense, several-month, in-the-field research journey that carried me in person from Hawaii to Canada, Belgium, and Colorado (and virtually from New Zealand to China), I can report that hemp is one of the most valuable crops for the USDA to encourage and subsidize with the greatest possible dispatch. As China is consciously on its way to doing, the United States should be cultivating two million acres of the stuff.

This recommendation isn't news to our leaders. In a 1994 executive order, President Clinton included hemp among "the essential agricultural products that should be stocked for defense preparedness."[13]

Whoever did the research for that wise conclusion had a lot of history to examine, starting long before modern researchers began testing hemp's stronger-than-steel fiber for use in body armor and aircraft components. Here's a portion of the USDA's 1895 *Yearbook of the United States Department of Agriculture*: "The hemp plant . . . has been so generally cultivated the world over as a cordage fiber that the value of all other fibers as to strength and durability is estimated by it . . . The plant is an admirable weed killer, and is sometimes employed . . . because it puts the soil in good condition . . . The value of hemp for fiber . . . and oil would seem to make its cultivation a very profitable one."

Actually, the White House researchers might have gone back much farther still. The Persians have called hemp *Shaah-daaneh*, or King of Seeds, for four millennia. I discovered this (before I fully realized what a concisely true name it is) when one of the engineers who was giving me a tour of the University of Manitoba's hemp-and-lime building insulation research projects in February 2013 mentioned offhandedly that our (quite technical) interview had brought back fond memories for him. Why? Because toasted hemp seed had "always been the go-to" after-school snack in his native Iran.

Absent Doritos and val-u-meals, Farhoud Delijani told me, hemp was a ubiquitous meal bridge for kids on the way to soccer practice. He couldn't wax wistfully enough on the Farsi treat of his youth. "Pop 'em in by the handful, shells on. Yum, it was just really tasty," Delijani told me, adding of the plant's now famous benefits, "We didn't know about the fatty acids, let alone the biofuel apps. It was just a very popular everyday treat."

That's a four-thousand-year-old message that should be heeded. Delijani mentioned biofuels. If you're like me, you're fairly desperate for a fossil energy replacement and you've liked what you've read about hemp's potential. But you've also wondered, "Really? The plant whose psychoactive side Abbie Hoffman thought should be mandatory before the stock exchange's opening bell is actually going to revitalize the economy and save the planet?" I'm excited to report that the answer is yes. Colorado biomass fuels consultant Agua Das and Colorado School of Mines chemical engineer Thomas B. Reed reported that an acre of hemp can produce power equivalent to a thousand gallons of gasoline.[14]

The hemp revolution is already under way, and, as with any new industry, its trajectory is market-driven. Today Canadian hemp farmers profit to the tune of $250 (U.S.[15]) per acre, compared with $30–100 for wheat. This on a crop that the Canadian Hemp Trade Alliance says will double in acreage to one hundred thousand acres by 2015 largely because 90 percent of it goes to the United States. We are, of course, the world's largest consumer of pretty much everything except affordable health care, and this includes hemp.

Consider these pages a playbook for the patriotic hemp farmer, entrepreneur, and investor who wants to help humanity transition smoothly from fossil fuels, tree farms, and monoculture. If you're simply hemp-curious, you'll hopefully finish this book a voracious consumer. That's a win for your health and the economy's. And if you're not an American, here's what Colorado rancher Bowman said is one of his motivations for establishing the plant's planetwide worth: "Family farmers like me are committing suicide in India and all over the world because the GMO cycle of debt is meeting climate change. I think we're going to find that hemp can help break that cycle."

American readers are about to notice that the rest of the industrialized world has a two-decade head start on the hemp revival. Embarrassing as that is, I've found that the slow start actually provides a slew of helpful lessons: We know what sustainably works (or can) in the marketplace. On our journey across four continents, we'll delve into the most promising real-world applications provided by a soil-stabilizing plant that can help replace or reverse three of my least favorite things: petroleum-based plastics, GMO monoculture, and environmental degradation.

CHAPTER ONE

Hemp Gets Out Ring Around the Collar

*B*efore getting to the industrial cannabis apps that have me so energized about my kids' kids' atmospheric prospects, I feel I have to address the Giggle Factor. Even though my family already sends Canada about three thousand dollars each year for products that start out life as Saskatchewanian hemp plants, after conducting all this research I'm myself still a little surprised at the extent of the plant's potential.

Sure, there's hemp in everything from my hand soap to the only diapers that hold up to brutal New Mexican line drying. These simply win in my home economy. But from there to "industrial cannabis can put the Coloradan and Kentuckian (and Ghanaian) small farmer back to work while forcing emergency board meetings at both ExxonMobil and Monsanto"?

When the folks who have been shouting about hemp's uses for decades might have come to their conclusions by what we can gently describe as anecdotal evidence, we somehow expect them to be wrong, or at least exaggerating. I think it might have to do with departmental tenure.

With apologies to the cynical, I can't help but report that I witnessed all of the cannabis uses we'll be examining. I saw half a dozen hemp-insulated houses. I studied multiple plans to power entire communities with plant fiber. I sent my holiday cards on hemp card stock. I even drove in a hemp-powered limo.

Bringing my belly into the research, I interviewed the authors of a new study showing that hemp-fed laying hens manage to pass on the plant's impressive essential fatty acid profile from their breakfasts into yours:[16] the eggs that you then eat. I even tried some. Fried.

The eggs were scrumptious. My fellow foodies will know that the deep apricot color in the yolks I enjoyed brings good news to the taste buds and life span. The two essential fatty acids we eat, omega-3 and omega-6, are aptly named, nutritively. They are essential. A balanced ratio of them is the modern Immortality Elixir. It's what Ponce de León would've brought home to the queen if he'd beached himself next to an Alaskan cod run. Sure, someday we may find it's all carcinogenic or a bunch of hooey, but here we are. I gulp the stuff.

Even though the facts on (or in) the ground convinced me that this one plant, properly utilized, can help form the backbone of a climate stabilization regimen while revitalizing the U.S. (and worldwide) small-farm economy and creating a community-based distributed energy marketplace, I still feel some hesitation alongside my excitement in relating them. That's because after twenty years as an investigative journalist who's covered some fairly incontrovertibly serious topics from Rwanda to Tajikistan, I can see the cynical media interviews on the horizon already: I frankly open myself up to Pollyanna accusations. Or Snoop Lion ones.

Even the most prominent hemp industrial player, David Bronner, CEO of the legendary Dr. Bronner's Magic Soap company and a fellow who was arrested in 2012 for tending a live hemp plant while both he and the plant were locked in a cage in front of the White House, told me he's careful in his interviews not to "overstate it."

A seasoned journalist, especially, doesn't want to come across as one of those *Hemp can do anything including get out Ring Around the Collar* people. But I have no choice: You're about to see that hemp's applications are as real as, well, Tide detergent. You actually can use Dr. Bronner's 18-in-1 Hemp Pure-Castile Soap to nontoxically and hypoallergenically clean your laundry. The gray water is so inert that you can water your garden with it while, yes, expunging that nasty Ring Around the Collar.

The 1970s commercial would at this point have the skeptical mom at the Laundromat ask, "Yes, but does it really work?" Two of Dr. Bronner's flavors (peppermint and baby mild) placed among the A-rated detergents in a 2012 field test conducted by the Environmental Working Group. Hemp's spectrum of industrial uses, I'm forced to report, are actual. *Broadly compelling* is the phrase that comes to mind.

With the exception of high-volume energy production (which is still largely in the drawing-board stage at the industrial level, although Czech hemp farmer Hanka Gabrielová is generating bio-energy from her harvest), all of the cannabis applications we'll be discussing have already been implemented in the real world. Which is to say that to accurately report on hemp, we don't have to rely on cheerleading hempsters.

Biologist Simon Potter can safely be described as cannabis agnostic. He works with hemp every day, at a Canadian government/industry joint venture called the Composites Innovation Centre (CIC) in Winnipeg. Short hair, buttondown style, British accent:

Simon Potter, Biologist, Sector Manager for Product Innovation, CIC

The forty-five-year-old Potter is excited to come to work every day, and this, of course, is a blessing for anyone. When Potter was giving me the tour of half a dozen hemp research projects at the Composites Innovation Centre (a government/industry joint venture in Winnipeg, Manitoba), his role was that of a generalist—he enthusiastically explained hemp insulation, hemp tractors, and hemp energy.

But as a biologist by degree and calling, Potter got particularly jazzed—perhaps a little pleasantly surprised—when I asked him about the microbiology of the plant. Especially the tiny battle going on during the part of the hemp-harvesting process known as retting.

Which is to say I never saw him so excited as when he smacked the poster-sized blowup of a microscopic green image like a headmaster and told me about the "fungal attack" he's trying to understand during this vital and risky part of the harvest process.

Retting is that fungal attack. And this is a good thing, because it softens the plant's outer hurd, or bark, which can then be removed for access to the gold within: the strong hemp cordage that already makes some, and in the near future is going to make so many more, of our industrial products. The retting process can take three weeks, leaving a valuable crop that's just survived a four-month growing season dangerously exposed to the elements.

Since he works for an organization that can afford what Monty Python called the "machine that goes ping," Potter can get 3-D views of what is essentially an in-the-field post-harvest microbiological battle that farmers have been intentionally harnessing for eight thousand years. "Once we understand retting, we hope we can make the process easier and more uniform," he told me optimistically as I stared at his chart. I'd never seen fungus so close up. It looked like a string of Tic Tacs.

In short, nature allows hemp's interior treasure box to open only at her own slow pace. It would be like drilling for natural gas, then having to simmer it outdoors for three weeks at a specific temperature before it could be used in a power plant or stove.

"We're also going to look at the internal chemistry and structure of the bast part of the hemp plant," he said of his next months' work. "This X-ray diffraction pattern here tells us a lot about where the strength of the fiber is going to be."

Which is all well and good for researcher Potter—I like to see a fellow who's living his dream. But what kept crossing my mind during my tour of the center's projects was this: As hemp is fully reintegrated into the world economy, which country has the advantage—the one putting its best biologists to work maximizing the plant's potential, or the one whose federal law enforcement bureaucrats are fighting tooth and nail to keep domestic farmers from even cultivating it?

Smart people are working on hemp, is the point I'm endeavoring to convey, and very few of them, as yet, are Americans. Potter seems quite willing to share the knowledge, though.

The guy's not an activist. Yet he, too, is bullish on the future of the industrial cannabis industry.

During a tour of the center's projects that included soundproof hemp walls, bash-proof hemp roofing, and cold-proof hemp insulation blankets (and, oh yeah, an entire tractor body made from hemp), Potter told me in the understatement of the day, "There's considerable interest in [industrial cannabis] materials." Shuffling through his desk, he added, "I have reports right here just on European, Australian, and Canadian textile applications . . . The economics of hemp have a very bright future."

CHAPTER TWO

Turning a Profit
Even with Medieval Harvesting Techniques

*H*emp, in its infancy as a modern industry, is already profitable in three distinct markets. In Canada it's seed oil, in China it's textiles, and in Europe it's construction (and other industrial markets). Massive demand is allowing a fledgling industry that has barely started identifying the kinks, let alone getting them out, to grow that 20 percent per year. I'll describe one "kink" right at the start to illustrate that hemp, despite its growing pains, nonetheless finds itself extremely lucrative at every stage from farm to supermarket shelf.

Our kink starts right in the field: The supposed weed isn't so easy to harvest. A seasoned, churchgoing Canadian farmer in coveralls named Grant Dyck, who grows hemp for what I can attest is his wife Colleen's delicious GORP energy bar company, told me as we crunched through his frozen back two hundred acres in the winter of 2013, "This is one of the more difficult crops I've ever worked with."

HEMP PIONEERS

Grant Dyck, Hemp Farmer

This is the guy actually doing it—has been for seven years. Today it might not impress a journalist to interview another GMO corn farmer about the latest developments in monoculture. But Dyck is one of only a few thousand North American hemp farmers at the moment. Certainly one of the few who wear Carhartts. So I think it's safe to say that the thirty-six-year-old lifelong Manitoban farmer was among the most valuable sources I interviewed for this project.

In the United States, everything related to hemp cultivation is weighted down with "What if?" It's the crystal-ball scenario in which pundits make their fortune. But Dyck partly makes his living from the crop, and so I listened up in his two-hundred-acre frozen hemp field as he stomped through months of white crust and handed me an armful of the previous season's harvest.

"Due diligence would be my first piece of advice," he said when I asked him what might save American farmers from some sleepless nights (or bankruptcy). "Do your research. Read all you can. Logistically, you've got timing issues. Adequate irrigation before planting is essential, for example."

He was just getting started. He's a very tall, clean-cut guy, and he was rubbing his forehead as he conjured up what looked like slightly stressful memories. "If you plant too early, you harvest when it's hot and wet—that's its own drying process. If you harvest too late, the plant's dry and you'll lose the seed as soon as the combine hits it. It requires good management to crop hemp." At least pesticides are a non-issue. What Dyck is trying to leave us with is the awareness that hemp might grow like a weed, but harvesting is a whole other ball game, like putting versus the long game in golf.

I wouldn't say that Dyck's in-the-field wisdom should bring every putative farmer down to earth. Excuse my generalization, but I think a third-generation North Dakotan is going to be able to handle any nuances required by a plant that has, after all, survived three-quarters of a century of taxpayer-funded eradication efforts along Nebraskan ditches. It's the eager-beaver newcomer who would be wise to do his or her due diligence. Professional farming is like professional anything. The marketplace (and, in this case, Mother Nature) will separate the wheat from the chaff. Oh, and in case farmers are wondering, Dick said you'll need between twenty-five and sixty pounds of seed per acre, depending on variety.

When I asked him why that is, he said, "After multiple harvests I'm still learning. My combine caught fire twice last year when the stalk got coiled around the blades."

Indeed, the overall cannabis-reaping process hasn't been much improved in the eight thousand years that humans have been working with the plant, though a lot of people, including a smart dude in Australia named Adrian Francis K. Clark, are working on it.

HEMP PIONEERS

Adrian Francis K. Clark, Inventor of a Hemp Decorticator

Sixty-six-year-old Clark is a citizen inventor in the tradition of John Harrison, the fellow who figured out how to measure longitude. What Clark is developing, though, doesn't have to do with geography and navigation. It has to do with that confounding post-hemp-harvest issue of retting—the removal of the plant's outer hurd, or bark, to get at the valuable bast fiber within. On the ground today, this is ol'-fashioned crop processing we're talking about—something out of the *Little House* books I read to my kids.

Here's how the province of Ontario explains the process to its cultivators: "Retting is the process of beginning to separate the bast fibres from the hurds or other plant tissues. It is done in the field, taking advantage of the natural elements of dew, rain and sun, or under controlled conditions using water, enzymes or chemicals. The method chosen depends on the end use to which the fibre will be put. Suitable industrial processes for water and chemical retting have not been developed."

Even though decorticators were being used as early as 1917, they either weren't ready for prime time or disappeared with the last iteration of the industry. In the digital age, farmers are often out there in the dew for weeks rotating the bundles of harvested hemp so they don't get too wet or too dry, and adjusting based on what Mother Nature has to say about it. Clark sniffs at this as a risky, potentially wasteful, and just downright passé mode. "Instead of letting the harvested hemp risk the elements, with the decorticator attachment on your combine, you can harvest the crop and remove the hurd on the same day. While the plant is still green."

Like the commoner Harrison's difficulty in getting Royal Society attention paid to his potentially world-changing invention, the Australian Clark has had some trouble making the case to the established hemp-fiber-harvesting authorities, particularly in China, which dominates the fiber market today at forty-five thousand tons of annual production.

"We've tried to keep the cost down even at the start, to $115,000 for a hand-fed model and $300,000 for a model you clip to the front of a combine," he told me via Skype (this compared with up to $10 million for commercial production models out of Europe, so Clark's real innovation might be relative affordability). "But in China cost isn't the problem. Officials told me I'd putting tens of thousands of people out of work—they sit in the field and peel the retted hemp [bark off] by hand. We are getting a lot of inquiries from Europe and Canada."

Out of the prototype stage and ready to manufacture and ship (Clark's Textile and Composites Industries YouTube videos are pretty amazing—you can sense farmers like Grant Dyck salivating as bast fiber shoots out the side of a combine during harvesting), what Clark and a few others are tackling is the first significant advancement in hemp processing since everyone in the Middle East got along.

"This is the key to the hemp industry taking off on a worldwide mass industrial scale," he told me. In a decade Colorado and Kentucky farmers might be thanking him, as well as making him wealthy, which is not a fate Harrison enjoyed: He died before collecting the Royal Society's twenty-thousand-pound prize for his longitude machine, the marine chronometer.

Clark's innovation is called a decorticator, and it solves the following problem: Today, if you want to exploit the plant's industrial fiber (and not just its edible seeds) when the crop is harvested six months after sowing, you first have to let it lie in the field for a few weeks. This is to allow the vital fungal soup we've discussed to soften up the tough outer husks. I can tell you from visiting hemp fields that these husks are better described as bark—they can be hard to snap at the base. And yet off they must come.

Too much moisture after harvest wrecks the crop, as does too little. There's not much you can do but turn the stalks like barbecuing drumsticks and hope the microorganisms do their job before the next front moves in. Learning this, I suddenly felt I better understood the plight of Russian serfs. One of those moments that makes you imagine life before Costco backup supply.

And yet for all its soon-to-be-upgraded legacy harvesttime kinks, hemp is economically viable enough that the Canadian industry is exploding. Those hundred thousand acres by 2015 still represent a small industry as a share of total Canuck agricultural acreage, but 20 percent is impressive annual sector growth by any standard.

A farmer who planted a thousand acres in 2012 netted $250,000. That's profit. And most of the half billion dollars that Canadian hemp generates in the United States comes from value-added final products like salad dressing and breakfast cereal. There's already hemp cereal at the International Space Station.

On top of that, the founder and president of Canada's biggest hemp seed oil processor, Shaun Crew, told me that a 20 percent annual growth figure is actually on the low end of what he expects at his Hemp Oil Canada into the foreseeable future. In fact, when we met in chilly Ste. Agathe, Manitoba, in February 2013, Crew was running around madly planning his latest factory expansion.

In the company conference room there was a photo of which Crew told me he's proud, as Hemp Oil Canada celebrates its fifteenth anniversary. It features him, hair less gray, posing with a member of the Royal Canadian Mounted Police (RCMP) while clutching a product called My Stash. It illustrates the law enforcement cooperation that Canada has enjoyed since the moment its industry revved back up in 1998.

Crew was there. After a few ups and downs that included a seed buyer bankruptcy in 2000 that hurt hundreds of farmers, he said the industry is now "taking on a life of its own."

"We just got approached by [food giant] Hain Celestial, and we're at capacity," he told me, waving toward the company's several dozen industrial presses in an adjoining building. Each of these squeezes out 350 green, lignan-thick gallons of nutritive hemp oil every day, working 24/7. "And we've already got a six-week delivery lag time now. You guys consume everything we've got."

By *you guys* he meant "you Americans," and he meant it indirectly: The company processes for wholesalers and does private-label packaging.

Hemp is harvested the way it was at the dawn of agriculture, and yet it's already creating millionaires. Crew related his stats and projections with a *Thanks for buying me the good seats at Winnipeg Jets games* smirk, Canada being a nation where the hemp oil flows like wine, and Americans will soon be buying a billion dollars' worth of it a year.[17]

Which is better than if we ate a billion in Canadian McNuggets, of course, since hemp is healthy for both our bodies and farmland soil. I checked out the Province of Ontario's official agricultural guide to hemp farmers. Right in table 1 under "Weed Control" it reads, "None needed. No herbicides registered."

CHAPTER THREE

Want to Make It in the Hemp Game?
Two Words:
Dual Cropping

I asked experts from four continents what advice they would give to nascent American hemp entrepreneurs, and while their tips were diverse and sometimes contradictory (they were, after all, advising future competitors), the phrase *dual cropping* invariably came up.

What it means is this: Hemp might be a miracle crop in its diversity of utility, but, particularly before economies of scale are reestablished in the United States, it is not necessarily a slam dunk in the marketplace in any one sector. So the digital age farmer/investor/industrialist needs to focus on two markets with one planting. At least.

To understand the three distinct product lines you're potentially utilizing with every hemp harvest, you have to know a little bit about the architecture of this remarkable plant. The triumvirate of usable parts (some of which we've already mentioned) are (1) the seed, (2) the bast (or high-quality long fiber), and (3) the shiv or hurd (often referred to as the woody core).

How dual cropping looks in the real world is something like this: Pick a hybrid cultivar (seed variety) that will provide both a seed harvest and a fiber harvest. Pretty simple.

How do you know which cultivar is right for your hilly section of Oregon, recent sugarcane plantation in Hawaii, or dusty former GMO cornfield in Wisconsin? When I asked Canadian/American hemp consultant Anndrea Hermann that question, the MA in hemp fiber agronomy spun a nearby globe and basically said, "Find a variety from your latitude, with your day length and your humidity." Day length, or photoperiod, is particularly important because the cannabis plant begins to flower as day length decreases. Certain cultivars will handle this more gracefully in certain latitudes. Premature flowering will reduce both a seed and a fiber crop's quality. Rainfall is another factor: During my research in Ireland, I learned that moist weather during harvest season has to be factored in—this might prove of use to farmers in a similarly wet place like Western Oregon.

In the Czech Republic, similarly, farmer and entrepreneur Gabrielová told me she's been growing the traditional European Carmagnola fiber cultivar, but is in contact with Finnish and Canadian seed providers as she searches for a dual-cropping variety that will provide both fiber and flowers for the health and beauty-care products she sells at her retail shop. "It's been fun finding out the long history of this plant in Czech culture," the 35-year-old told me via Skype. "It's in our folklore. We have towns and even birds who like to eat the seeds named after hemp."

Even if you're not lucky enough to live in places like Nebraska, where as we've discussed natural selection has chosen the variety (it's that one you find in your irrigation ditch), it's still a matter of easy if methodical research to find cultivars well matched to your ecosystem. Unless, of course, you live in unusually chilly places like Antarctica or Congress.

A good starting point is the online cultivar list that Canadian federal government provides its hemp farmers.[18] Or the farmer guide that the Brits who make hemp-and-lime building materials distribute gratis.[19] From there, delve deeper into a prospective cultivar's climate of origin, then call some farmers there to see if the climate, soil, and overall agricultural season sound similar to yours. Do your advance work, in other words, as you would when starting any business.

One cultivar Hermann suggested for dual cropping in the northern United States is called Alyssa (it was the subject of her graduate work). Come harvesttime, the farmer can sell her seed to the oil processor (or better yet, unite with her neighbors to own their own processor), and then sell the fiber to, say, a construction materials entity like the already established American Lime Technology, for use in building carbon-negative homes (more about those in a little while).

That isn't even the most exciting part. This is the most exciting part, which I'm going to suggest even though one hemp expert called it a "current dream but a possible future if the numbers add up": Imagine squeezing in a third harvest from the tough remaining hurd of the plant, to use for something fairly important: putting Chevron out of business—or at least forcing it to radically change its business model, thus ending a lot of U.S. and local teenager deaths in the Middle East.

The fact is, I found that planet-friendly "biomass energy production" is already happening. There are, for example, Austrian farmers generating their own (and some of their local grid) power from farm-side, cell-phone-monitored personal power plants. The power comes from stuffing waste farm fiber into a truck-sized device's specialized double furnace chamber, creating energy via a relatively clean, anaerobic process called gasification. We'll be talking more about this later, but for now know that those thousand gallons of gasoline power that Das and Reed tell us an acre of hemp can produce comes via gasification.[20]

Gasification (or more generally "biomass combustion") has not been implemented on a large scale with hemp to date, and one European hemp consultant cautioned me that cannabis hurd doesn't provide as efficient a per-acre energy yield in a biomass combustion application as some other crop waste. But, heck, if you have it lying around by the ton? I mean, for a decade and a half the Canadians have just been burning it in the field.[21]

Props to the Canadians for jump-starting the industry, but one day, single-market hemp farming will be considered laughably wasteful. Indeed, with the amount of research, brainpower, and private and government support industrial cannabis is generating everywhere in the industrialized world except the United States, no doubt new applications will emerge for all parts of the King of Seeds plant in coming months and years.

Which is to say I'm nearly certain that when I update this book in five years I'll have to slap my forehead and apologize for missing hemp's explosion into, I don't know, water purification, malaria antidote making, cold fusion, computer chip multitasking, or Roger Ailes humanizing. I know of one fellow who's developing a hemp plastic bottle to replace our landfill-clogging petroleum-derived bottles. He's raised seventeen thousand dollars via Kickstarter. Another hemp entrepreneur, Wisconsin's Ken Anderson, is delving into the federal approval process for hemp-based highway soundproofing material. "It absorbs twice the sound that concrete does," he told me.

But I can only report on what I've seen today, and in the coming chapters we'll look at some major hemp sectors to get us started dual cropping a plant from which China today nets close to two hundred million dollars, according to Shangnan Shui, a commodity specialist at the Food and Agriculture Organization of the UN. Instead of restricting and threatening its farmers, China sends its president to tour hemp research facilities.[22] That's a multibillion-dollar policy difference.

Now, you've got to start somewhere. If you're an American farmer (or processor) who wants her work to pay the mortgage, you're asking, "In which industrial cannabis market should I begin when Congress finally gets around to passing the FARRM Bill?" The answer to that question is unequivocally "the seed."

This is actually two markets: seed oil and the accompanying high-protein "cake" that's left in the press after extraction, the world supply of which is currently snapped up for high-value food and body care products.

The reason to start with the seed is that the template is already in place: If you grow it, there's a buyer for it. In fact, it's pretty much the only existing North American market on the production side as the young industry reaches its fifteenth season since Maple Leaf re-legalization of the crop. Nearly all of Canada's considerable government support of its lucrative young industry has gone into developing hemp varieties that create goopy, omega-rich food oil.

Many farmers I spoke with grow a cultivar called Finola, which is so short-stalked and seed-dense you couldn't use it for fiber even if you wanted to. This is how they've liked it up there up until now.

HEMP PIONEERS

Colleen Dyck, Founder and President, GORP Clean Energy Bars

Thirty-six-year-old Dyck is one of those pint-sized wonder women who seem to have plenty of time to pick the mint for the tea she's brewing for you despite (a) training for a triathlon, (b) chauffeuring four young children to hockey practice and church groups, and, oh yeah, (c) launching a literally homegrown energy bar company in her basement, which is where I visited and snacked with her.

As I sampled a cocoa/hemp/almond flavor (she's proud that the packaging is resealable, since the bar, nutritionally, is "more than a small snack"), Dyck told me that "educating people about where their food comes from" is part of her calling. "If this venture helps people get in touch with the story of their daily food and how it relates to their life, I feel like it's a good thing. That's one reason our farm's hemp and sunflowers are in our products." Watch out, Power Bar.

The world probably wouldn't be blessed with GORP Clean Energy Bar company had the diminutive, raven-haired Dyck not won "fifteen thousand dollars I did not have" in a competition called The Great Manitoba Food Fight in 2009. "It takes so much money for R&D,

shelf-life testing and mold testing, and everything else you have to do to bring something to market."

When I asked her whether it's a bigger accomplishment to launch a business or raise four kids, she laughed but didn't hesitate. "Raise four kids."

So if it's your dream to make a hemp-seed-based energy bar, a healthy juice (I'd personally love to see a mango/hemp/chia blend), or a Hemperific Hydration skin cream, you're well positioned.

A hemp wedding gown? Not so much, unless you're fluent in Mandarin and prepared to commit some serious industrial espionage by getting a job as janitor at Beijing's Hemp Research Centre. Chinese industry is a decade ahead in the "cottonization" of the plant's long outer bast fibers. In the industry the softness of the finished textile is referred to as its "hand." This can be read as "its readiness for the shelves at the Gap."

Attractive as replacing cotton with hemp sounds when you know about the former's monster water needs and obscene pesticide footprint (cotton uses 25 percent of the world's pesticides, according to the Pesticide Action Network), several industry players cautioned that re-adding chemical processes to natural fibers kind of defeats the purpose.

As someone who's written a book about living petroleum-free and who has yet to feed his children (or clothe them in) much that is non-organic, I of course powerfully agree. As a cannabis journalist who's just talked to two dozen experts who work with the plant every day, I'm convinced that the additives and binding agents that digital age applications demand—be they for a bathrobe or an aircraft's door panel—can themselves be best derived from nontoxic (often plant-based) sources. Accordingly, a Portland Oregon-based company called Naturally Advanced Technologies offers a sustainable enzymatic fiber softening process called Crailar that the company website says "drastically reduces" chemical and water usage.

If hemp bathrobes are your dream application, you're truly going to have to be a modern hemp pioneer, growing (or if you're a manufacturer, contracting with farmers to grow) known textile varieties of the plant that put the *dual* in your dual cropping. Then you'll have to borrow or innovate technologies for fiber softening. Good luck with that: I'll buy the plush American-grown bathrobe. I'd like mine deep-piled, please. Like sheepskin, or polar bear fur. It gets cold under the stars on my ranch at fifty-seven hundred feet.

So all of today's hemp industrialists tell me to tell you to start with seed applications. Find the cultivar that works in your latitude per Hermann's and the Canadian government's advice (hug a publicly funded cannabis cultivar researcher, people). And if you and your friends have a spare couple of million clams, start a cooperative processing operation. It's not that complicated to make the oil and the cake once you've harvested your eight hundred pounds of seed per acre.[23]

There's a reason they call it an oil press. As Hermann explained it (and in fact demonstrated to me in Winnipeg), you just smoosh the seed and the oil oozes out. It's a refreshingly mechanical technology, unless you count the newfangled fringes Hemp Oil Canada is now offering, like automated bottle labeling and expiration dating. This

HEMP PIONEERS

Barbara Filippone, President of EnviroTextiles

Based in Glenwood, Colorado, EnviroTextiles is a woman-owned industrial hemp and natural fiber manufacturing company that has offices in three countries, supplied President Obama with a hemp reelection scarf, and is the largest importer of hemp textiles in the United States. But what Filippone told me breathlessly was the reason she was contacting me is that her company, which is worth fifteen million dollars, had just earned approval from the USDA BioPreferred program, which promotes the purchase and use of biobased products. "Hemp's already creating jobs while still a schedule one narcotic," she glowed.

The victory was a long time coming. "I worked in China for nineteen years and for thirty-seven in plant fiber, as well as three years qualifying to be a government supplier," she said. But with a deep belief that natural fibers are the only future option with petroleum-based synthetics on the endangered list, she was confident hemp would win out in the end.

"The federal government knows hemp is an alternative to cotton that's drought-resistant. The military knows it—I've been speaking with them. Cotton's done. China knows it, too."

Filippone gave a very bottom-line reason China is moving away from cotton. "It uses too much water and pesticides," she said. "They have no choice."

Since she is so seasoned in the real-world economy, Filippone is not shy about offering advice to the tide of newcomers who are already becoming both her colleagues and her competitors.

"It's time to learn how to be a real businessperson," she said with a touch of the scalded tone that I recognized from my grandfather's admonitions about the business world. A competitive landscape is going to "eat" the naive, she added. But she also offered assistance. "I've been around the block many times. I can help those who want government approval in other hemp niches, including construction."

dang nontoxic bioproduct goes rancid after a year or so. Almost makes me wistful for the time society thought long shelf life was a good thing.

Now, plenty of "in" superfoods have enjoyed their fifteen minutes and have then disappeared from shelves (seen much açaí lately?). But—and don't take this as investment advice, I'm just typing here, albeit after a lot of research—I don't see hemp going away. Ever.

As we sipped hemp coffee in his hemp-stalk-enshrouded executive office, I asked the seed-processing mogul-in-the-making Shaun Crew what he will do when American cannabis prohibition ends.

He leaned back in his chair, inhaled deeply, and proved quite ready with an answer. "The moment your guys' drug policy changes, we'll parachute a facility exactly like this, turnkey, into Grand Forks,

North Dakota. Or maybe Fargo. There's a lot of inexpensive acreage to grow on down your way. And massively growing demand. It'd be an easy move for us."

He said this in a very relaxed tone of voice. Kind of made me believe his characterization of this factory replication project as a manageable, low-multimillion-dollar investment.

In addition to letting me know how ready the American market is, Crew gave me another vital (and to me surprising) piece of advice. When I asked him about a hypothetical "hundred-acre" American dual-cropping hemp farmer's prospects, his arm shot up in the universal *Talk to the hand* gesture.

"Stop right there," he said emphatically. "You'll want to start with one thousand to three thousand acres to make it viable. Otherwise it's a hobby farm."

Yeesh, I thought. *I raise a few goats on forty-one acres and I feel like Ted Turner*. But the guy does work with forty farmers and have twenty-one employees, so I did the math: Cultivating two thousand acres at $250-per-acre profit, you are, to quote Eddie Murphy's Billy Ray Valentine in *Trading Places*, buying your kid "the GI Joe with the kung fu grip" for his birthday. Scale, in hemp farming, helps a lot.

Crew offered his last tip of that informative morning almost as an afterthought as I was leaving his office, insufficiently bundled for Manitoba in February. It involved value-added marketing. "Your oil's value triples once you put it in, say, a skin care bottle," he said. Be a farmer, in other words, but be more than a farmer. Like the Dycks with their GORP energy bar. Turned out to be a piece of advice to which Colorado farmers and hempreneurs in particular are paying a lot of attention.

So I think we've established that exploiting the high-lignan hemp seed oil is an obvious and relatively safe place for American hemp

HEMP PIONEERS

Shaun Crew,
Founder and President,
Hemp Oil Canada

I don't know if he'd consider himself the J. P. Morgan of hemp seed oil, but the fifty-four-year-old has been in the modern industry since day one, in 1998. Actually, since a month before Canadian hemp re-legalized hemp that year. His company, along with nearby Manitoba Harvest, is a major player in Canadian seed oil processing, he can hardly keep up with its growth, and he openly cannot wait to be able to contract with North Dakotan farmers a few miles to the south—he needs the acreage planted to meet seed oil demand. When we met at company headquarters in Ste. Agathe, Manitoba, Crew had come straight from the airport after keynoting a European hemp conference. And yet, as with several other hempreneurs I met, what struck me the most about the guy is the fun-loving attitude he brings to his work.

For one thing, he sits at an executive desk framed by a version of Grant Wood's *American Gothic* portrait where the somber farmers have a hemp leaf stuck on their pitchfork. I guess I shouldn't have been surprised. Crew freely disclosed to me that, back when hemp was far from a sure bet, he channeled his entrepreneurial drive by taking a risk on the plant at least partly because of his "familiarity with and friendliness to" all forms of the cannabis plant.

This stayed with me, as some hempsters try to separate the non-psychoactive uses of the plant from the social/medicinal side. Yet one of the most successful ones didn't seem to think that appreciating all uses was something to hide from an author interviewing him on the record. Crew also wanted to talk hockey, but that was just a Canadian thing, not a cannabis thing.

Simply liking the plant and what it obviously offers humanity's economy and spirit is what's behind the *We win!* smirk he's wearing in his picture alongside a red-jacketed RCMP officer while holding a bag of his first approved hemp seed product, My Stash.

None of this means that Crew doesn't take his work seriously. He's a workaholic who doesn't notice the hours he puts in—he told me that even with yet another Hemp Oil Canada factory expansion under way, he usually calls it a day at least in time to hit his family's Winnipeg Jets seats by the start of the second period. And he's another hemp pioneer who's happy to offer advice to putative American hemp entrepreneurs.

"Adopt a *Walk before you run* attitude," he advised. "Develop a market for your products and the processes by which you're going to operate. The pot of gold at the end of rainbow comes after a lot of work and, in the U.S. this will especially be true, market development."

pioneers to start. When it comes to the likely growth curve of the seed market, it seems like there's no end in rapacious sight. Not for me, that's for sure. Right now I'm sipping my usual morning shake full of blueberries, bee pollen, and hemp seed oil. This last ingredient is produced by the Nutiva company to which we've been introduced from Canadian-grown (dang it!) hemp seed. Chances are I walked past the press that processed it in 2013 while researching this book.

So if you want to make me and millions of other eager-to-be-locavore consumers happy and at the same time develop a business that'll put you in the good hockey seats, you can emulate the Hemp Oil Canada model: Work toward vertical control of your regional industry by teaming up with half a dozen (or many more) other farmers and investors on a "turnkey" seed-oil-processing facility yourselves. Market it as a hemp microbrew.

It's not like the oil press machinery is hard to find. Nor consultant advice even for farmers in the field. The whole time I interviewed consultant Hermann in Canada, her phone kept ringing. I heard her tell clients in places like Quebec and South Africa things like, "Well, sure, I'd recommend straight-cut combining at reduced speeds—you can lose yield with swathing."

Though you better start drafting your business plan now: Crew already has a Hemp Oil USA logo. This is not a fellow who, after fifteen years building his company as a trailblazing entrepreneur, chases wild goose markets. The Canadian economic imperialist knows as well as anyone in the world that the American hemp appetite is insatiable, and that North Dakota, Colorado, Kentucky, and much of the heartland is ready the minute U.S. federal policy finally surrenders to inevitable market forces. "Hemp hemp hooray!" were Crew's final words to me.

CHAPTER FOUR

Grow Your Next House
(or Factory or Office or
High-Rise or School)

M y fingers were numb from tapping notes into my
phone in the subzero Canadian heartland provinces
late in the morning of February 22, 2013. The primary thought I was trying to record midway into this, my third
hemp building site visit in as many days, was, *Okay, you're an
American hemp businesswoman paying the mortgage with hemp seed
oil toothpaste grown and processed nearby. What can you do with the
fiber to maybe open yourself a savings account?*

I see now that through the frost-induced typos and misguided
autocorrects I also managed to record the secondary thought, *Wow.
Manitoba. What a difference an inch makes.* I was thinking of nearby
North Dakota.

Specifically, I was trying to map the conversation I had just
enjoyed with the Province of Manitoba deputy minister for housing
and community development Joy Cramer onto her American coun-
terpart sixty-five miles south. That'd probably be North Dakota

agriculture commissioner Doug Goehring, since the Roughrider State doesn't have a housing secretary.

The twenty-below morning would be familiar to both politicians, as would the rate of neighborhood vehicles gliding gracefully into snowbanks per hour (it would soon happen to me), and probably also the truly inspiring organic cinnamon buns. Closely observing local winter survival techniques, I lived on those warm, caramel-dripping sponges during my Canadian research.

What would be different those sixty-five miles to the south was the likely federal drug enforcement authorities' reaction to Cramer the Canadian's casual statement to me that she "just wants to help our local hemp farmers." At least as of 2013.

See, the materials used to insulate the half-completed house inside of which we were then standing were cannabis plants grown a few miles away in the Manitoba countryside. Here in downtown Winnipeg's upscale "granola belt" neighborhood where she herself has lived, Deputy Minister Cramer had been telling me about the provincially funded hemp house project, after we'd clumped up the icy porch stairs to the homesite.

"We're particularly interested in this project because the insulation and wall materials are being built of hemp and lime, and if it performs, it's another market for our farmers and our construction industry," said Cramer, as polished a politician as I've ever met. "We know it's a big crop for Canada and we're proud that our Manitoba hemp farmers are some of the biggest exporters in the country. We have high hopes for our hemp industry here."

Brian Pollack, the project's contractor, demonstrated the tamping of the "hempcrete" insulation on which the house is relying. He told me hempcrete is essentially a self-curing and breathable mixture of dried cannabis hurd and lime, absent concrete's toxic ickiness and high heat needs. "It comes out of the mixer light and airy," he said, and though there's something of a science to getting the material ratios right, once you do, "It's easy to use and fast to build with."

When she said that provincial deciders are watching how the hemp house "performs," Cramer told me she meant in three categories: (1) cost competitiveness versus standard methods in the box-store era, (2) ease of replication in remote areas of the province, and, (3) (obviously a vital one at twenty below in February) how it performs thermally through a Canadian year.

This last category includes not just energy costs in a place that endures drastic annual temperature and humidity swings, but also how comfortable the not-yet-selected family that will move in by the end of 2013 will be. Cramer—a senior enough official to have a driver and pull off a beaver cap in a casual interview with a shaggy American author and filmmaker—told me her office will be closely studying the reports that come in surrounding "air quality, mold control, and, of course, warmth."

Yes, this house is going to be Home Sweet Home for real tenants, probably by the time you're reading these words. Hearing this, I gave Goehring, North Dakota's agriculture commish, a buzz to see how it feels for North Dakota farmers not to receive Cramer's kind of support (let alone $250-per-acre profit).

"It makes you some days want to shake somebody," the Republican said. But he said it in a very polite prairie way, as though he was going to follow up with, *Did you hear Jennifer Svenson is running for deacon?*" He reminded me of Garrison Keillor's more deliberate cousin.

"We have producers who would like to grow industrial hemp," he continued. "They know that state law allows them to be licensed, but they also know that they then have to apply to the DEA to be allowed to raise hemp. We in North Dakota asked the DEA to waive that registration. They denied us that.[24] It's a bit disingenuous for the federal government to . . . completely ignore the great attributes of industrial hemp. It should be utilized as a commodity for the public good."

I asked Goehring if he knew that Canadian farmers were banking 250 C-notes an acre within shouting distance of the border. He did. "And that's the other thing about it," he said, still politely soft-spoken, but getting what no doubt passes in the northern heartland for worked up. "Agronomically, hemp works in our soils. It can be another in our healthy crop rotation system."

At last Goehring really exploded into what a New Yorker would call a soft speaking voice. "It's amazing. We have farmers who live near the border, and six miles away cultivators are growing this very viable product. For those who have expressed interest, it's a lost opportunity for a fine product that's utilized all over the United States."

I hung up the phone wondering, *What does it mean that I seem to be agreeing with a lot of Republicans on drug policy lately?* Then I remembered my mission—the thought that I had used most of my five seconds of exposed Manitoba finger blood flow to tap into my phone: Where might the first hemp fiber killer app reside? I was about to find out I had been looking at it while interviewing Deputy Minister Cramer and dreaming of cinnamon buns.

If you're like me (and Deputy Minister Cramer), you're wondering if we know how hemp/lime insulation performs in an actual finished house, or even what this hempcrete insulation is. Starting with the second question, I looked for a scientific description of what makes fluffy hempcrete the effective insulator that the Manitoba builders were telling me it is.

The explanation came care of Farhoud Delijani, PhD student in biosystems engineering at the University of Manitoba's frankly awesome Alternative Village. You'll remember him as the fellow who ate all the King of Seeds as a kid in Iran. He showed me several underway hempcrete experiments, including one surrounding energy efficiency. I got to touch the stuff, and it is, as advertised, "light and airy." You can see the hemp hurd in it even after it's cured into rock-solid blocks. It looks like the shredded stuff people toss into guinea pig cages.

Turns out that when it comes to insulation, you want to trap air (like a wet suit traps water), so "light and airy" material is not just easier to work with than fiberglass, it's an essential part of the formula. Delijani told me that hempcrete in twelve-inch-thick sections has an R-value (this is how insulation is rated) of twenty. That's extremely competitive, even without factoring in the absence of material toxicity and the centuries-long durability that home designers insist you'll get in your hemp house.

What's the bottom line performance of hemp as an insulator? "The study is still under way," Delijani said. "But it clearly takes less energy than the control group to keep a house heated to twenty-one degrees Celsius [about seventy degrees Fahrenheit] throughout the [Canadian] winter." That was music to the ears of a solar-powered goat herder rooting for his two young kids to have a livable planet.

And hempcrete is very easy to apply on-site, Delijani and others told me. Mechanically, what the builder is doing during hempcrete construction is described by the website HempBuilding.com this way: "The hempcrete is cast around a timber framework. This is achieved by tamping down between shuttering, or it can be sprayed against a formwork . . . The hempcrete [is then] finished with a natural paint." Some companies even sell complete construction blocks with the hempcrete already inside.

Now that we can visualize what hempcrete is, we can move on to our other question, that of finished hemp structure performance. To do that, we leave the lab and head into the real world.

We've all seen promising wonder technologies that flop in the marketplace for one reason or another (cough cough Bill Gates/ the Chinese/Walmart offer a watered-down version cheaper). In fact, when I wrote for a computer magazine just out of college, I learned that there's a name for such technologies: vaporware. An actual working product that lives only in the hopes of the company's founding team and perhaps a harried publicist.

So I set my Anti-Vaporware Detection Meter to "sensitive" and reached out in an effort to learn how hempcrete performs in a house that's faced the elements. I found the answer in Dixie by way of New Zealand. Greg Flavall is co-founder of a company called Hemp Technologies, which built a hemp house in Asheville, North Carolina.

I watched a comprehensive video about the house. It furthers Delijani's description of what hempcrete is, even showing the schedule-one-narcotic hurd before it goes into the mixture. The result performs better than pink fiberglass insulation, even in sections that were left exposed to the elements for a winter. The Twenty-First Century Construction Revolution is hereby televised. Google "hemp technologies asheville north carolina house." It could be a turning point in the human pursuit of shelter since the questionable decision to emerge from the cave.

When that Asheville project was finished in 2010, Flavall moved the heck back to his native North Island of New Zealand largely because, being such a believer in hemp building technology, he wanted to be closer to the one key factor keeping it from being cost-competitive Stateside today: legal local sources of hemp.

I'm no architect, but Flavall's North Carolina house project just looks to me like a nice big American house. I found the "carbon-negative" and low lifetime energy use claims in the Hemp Technologies video compelling enough, though, to Skype with Flavall in New Plymouth, New Zealand, at some unusual hour. I think we were in fact on different days.

First off, Flavall stopped me right away when I asked him when he thought hemp construction would be cost-competitive. I posited that one first because I knew that, even if hempcrete cures cancer, a certain type of reader is only going to want to know what it costs per square foot to build.

"It already is cost-competitive," the fifty-four-year-old said in his Kiwi accent. "That house in Asheville was the first one we built, it was in a high-wind zone, which meant we needed to bring in additional structural engineers, and we still finished it at 150 dollars a square foot not counting grounds work and landscaping. That compares favorably with regular construction."

Okay, that emboldened me to ask about these carbon-negative claims. "The house eats carbon?" I asked. "Just cleans the atmosphere when you're sitting in the living room doing a jigsaw puzzle with the fam?"

"It does," Flavall said in a tone I'd describe as calm confidence. "First off, it's 80 percent more energy-efficient than a regular building—it costs twenty-five cents a square foot per month to heat and cool—which is a testament to the quality of hemp as a thermodynamic insulator. But in addition, the lime feeds on carbon dioxide as it [the lime] hardens over the course of years. It wants to go back to rock so it absorbs carbon from the air while making the house stronger. That house is going to last hundreds of years."

"What about the houseplants?"

"Excuse me?"

"Don't the houseplants need carbon dioxide?" I asked. "Does the lime steal it from them? People want to have houseplants."

"The houseplants are fine," Flavall assured me. "They just provide another carbon sink."

Flavall also said the house was getting about the same 2.4-per-inch R-value from hemp walls that the Alternative Village studies were showing at the University of Manitoba. "That more than meets building codes in America," he told me from a distance of seventy-seven hundred miles from the drug war.

Tim Callahan, who designed another North Carolina hemp house, also said that significantly lower labor costs associated with hemp building are a major factor. "It's doable with not that large of an investment, especially once we have repeatability."

Sounded terrific, but Flavall and Callahan don't have to live in their creations. Amazingly, it seems there's yet another house in Asheville (talk about the New South!) whose occupants seem quite satisfied with the cost and comfort level in their hemp home.

"This is the . . . future," owner Karon Korp told CNN, while her husband, former Asheville mayor Russ Martin, raved about the low utility bills he pays in the upscale, thirty-three-hundred-square-foot pad. A clean interior environment was the big selling point for the family. "The house itself is an air filter," said its designer, Anthony Brenner.

Flavall's Hemp Technologies website even has a page of prototype hemp structures they'll build for you, ranging from a single-family loft to an apartment complex to, for gosh sakes, a mixed-use commercial site.

Still think it's a pipe dream? Hempcrete is featured in a new, 195,000-square-foot British Marks & Spencer department store. That project delivered me a memo: *Check out what's happening with hemp across the Pond.*

I guess that North Carolina's mini hemp housing boom answers the livability question. I mean, I didn't have the budget to time-travel ahead three hundred years to see if all of today's hemp houses are still standing and happily occupied as promised. But the revolution well under way in European construction—which was the biggest stunner in my research for this project—is kind of like time travel. Two decades' worth.

Which is to say that all this North American V8 head slapping about hemp's proving a serious contender for "Go-To Building Material of the Sustainability Era" is old hat on the other side of the Atlantic. In fact, hemp mixtures are being used not just for insulation, but for load-bearing-construction applications as well.

No-brainer was Lime Technology vice president and director of technology Ian Pritchett's choice of phrase about hempcrete[25] building when we Skyped and I asked, "Why the heck aren't you Europeans telling us about all this hemp construction?" Like most of the hemp executives with whom I met, Pritchett was not wearing a tie when we interviewed, even though his construction company is already classified, thanks to its ten million euros in sales in 2012, as a British SME (small to medium enterprise). They're past the start-up phase, is what I'm saying. This is not a fellow hawking granola at a crunchy trade show.

Even with two hundred houses and fifty commercial buildings under his belt, including an attractive little English housing complex known as The Triangle, the British hemp builder did attach a condition to his no-brainer assessment.

"The target market for us is people who are building a structure they'll run themselves," Pritchett told me. "So that the construction costs and the energy costs are coming from the same person. Then it's a no-brainer. Even when it's a little more expensive to build, it's much cheaper to run. But if the builder only cares about the immediate costs, then that's for now a tougher sell."

The longer view Pritchett's speaking about starts with climate change mitigation at the very building site: You have to heat concrete as high as three thousand degrees. Not so hemp/lime. This is why many of the conclusions you'll find in the European Industrial Hemp Association trade group's "Assessment of Life Cycle Studies on Hemp Fibre Applications" paper[26] show hemp pulling ahead— or farther ahead—of a diverse array of synthetic and fossil-based industrial applications once length of use is factored in.

Seeming to anticipate hemp building similarly proving a no-brainer for northern Canada back at the brisk Winnipeg hemp house site, Deputy Housing Minister Cramer called hempcrete's likely performance the "easy part" of the project. On durability alone, she echoed (albeit in a sub-Arctic version) what hemp-friendly Hawaiian legislators told me about the plant's impressive resistance to the tropical plague of termites. "Here the problem is mold, and it's doing great in tests at the university [of Manitoba] so far," Cramer said.

The actual trickiest hurdle that future builders and entrepreneurs will face, Cramer surprised me by revealing, is "each municipality's unique building codes. We're documenting how we maneuver through that whole process, and our construction industry players are watching. City bylaws are always a pain for builders."

Cramer might be ironing out the mundane wrinkles in construction bureaucracy, which I'm sure is an essential part of any building equation when a new material enters the marketplace. But it struck me by the time I started skidding back out into Winnipeg traffic that day that what we're talking about with hemp-based construction is more than a revolution in the building industry. It's a revolution in society.

Why do I say something so dramatic? Because cement plants alone contribute 5 percent of global carbon dioxide emissions, and the construction industry was responsible for more than 8 percent of the U.S. GDP in 2007. Make that huge industry not just sustainable but domestically produced and you're at once betting on America's economic and atmospheric future.

North Carolina hemp house designer Callahan told me that the only thing keeping hemp building technology from domestic cost-effectiveness today is having to import the actual hemp.

"That would be a huge boon to us," he told me of being able to get ready-to-go building material from Cleveland or Lexington, Kentucky, instead of the UK or Canada. "And it would almost certainly be the turning point for the domestic hemp construction industry."

Construction, in other words, is going to be the first domestic hemp fiber breakout market. Steve Levine, CFO of the Hemp Industries Association trade group and a fellow who's been selling hemp products in the United States since 1997, said he has little doubt that hempcrete will be the first dual-cropping sector to explode.

"If I were a venture capitalist with ten million in play, I'd invest in building materials," he said. "Once there are processing plants Stateside, once Kentucky and Southern California are growing industrial cannabis, the battle is mostly won and we'll see exponential growth."

CHAPTER FIVE

Heck, Grow Your Whole Tractor Out of Hemp

Y ou know James Bond's gadget guy, Q? Best job in the world, as far as I and about ten million other fourteen-year-olds-at-heart are concerned. That's pretty much Simon Potter's gig at the Winnipeg-based Composites Innovation Centre (CIC) we've been visiting. His title is sector manager for product innovation. Oh, how I crave such a title. "Hold my calls, Ms. Moneypenny, I'm working on the Invisibility Suit this morning."

As we toured the warehouse-sized CIC labs (hidden in a nondescript outskirt like British Intelligence headquarters would be), I kept expecting to see wristwatches shooting poison darts at targets and men in beekeeper suits absorbing small rocket attacks without harm. There was even a Moneypenny-type character at the sleek, overlarge, semicircular front desk, who offered me a kind of futuristically labeled water bottle.

Only Potter doesn't work for MI5. He works for the future of the atmosphere. Funded, of course (since this is anywhere on the planet but today's United States), as a nonprofit by the federal and provincial Canadian governments in partnership with various private foundations and industry players.

Hemp is prominent here at this facility whose formidable engineering and biological minds are dedicated to designing not the cheapest, not the most appealing-to-young-demographics, but rather the *best* of tomorrow's industrial materials.

And the multi-team work at the center is showing that biocomposites—naturally sourced plastics and replacements for toxic or petroleum-dependent materials like fiberglass, particleboard, and plywood—are performing best in an incredibly broad array of industrial applications.

"I don't know why we forgot, institutionally, about this plant's uses," Potter said of cannabis at one point on our tour. This is a well-funded scientist who can, and does, work with any material he chooses.

Why is the remembering currently under way so vital? Composites are the fastest-growing segment of the wood products and plastics industries.[27] An eighty-billion-dollar market, Potter told me, rapping his knuckles on what looked like a huge, shiny vehicle hood. "And we can replace 30 percent of it immediately with biocomposites like this hemp tractor hood."

I stopped in my tracks. Yes, what really blew the mind at the CIC was much closer to home than an Invisibility Suit for this solar-powered goat herder. I realized immediately that I'm a fellow who's in the market for a sustainable tractor. I just hadn't realized it was an option. I've been using a machete for my squash.

The tractor hood was shiny, curved in an appealingly contemporary design arc, and undentable by my hardest palm-heel hammer punch. It was beautiful. And it was grown from the local hemp harvest.

Plants, I had previously thought, go in the garden. They don't go into the manufacture of heavy machinery. To me, what I was looking at was as strange as a space shuttle or a senator being made out of industrial cannabis.

"It'll make a lighter, stronger, and considerably more fuel-efficient vehicle than the usual composites that go into heavy industrial structural components today," Potter told me. "And it's grown by the farmers from the material it's going to harvest."

That's not just a cute locavore gesture. "Natural fibers are cheaper than synthetic fibers, to start," Potter explained. "And this tractor's body embodies a lot less energy in production than synthetic fiberglass body components." Meaning, he said, "It off-gases far less carbon in production. Fiberglass is an energy-intensive process to make."

"Why do you think The Hemp Reaper or whatever you're calling it will result in greater fuel efficiency?" I asked.

Potter pointed at a nearby hunk of "traditional" plastic. "Because it's 30 percent lighter than that synthetic composite. That will translate to greatly increased fuel efficiency in the vehicle." (Also, as we'll see, it can one day be fueled by hemp.)

Then he laid the zinger on me when I asked as skeptically as I could, "Are we really going to see hemp semi trucks? Hemp airplanes? Hemp as a major industrial component competing with or even replacing the major materials of today like steel, fiberglass, and petroleum-based plastics?"

"I think it's absolutely inevitable," Potter said confidently and without pause. "It's the only way we're going to have structural materials in the future. We can't rely on fossil sources anymore."

The smart men and women at the vanguard of biocomposite research are on the case. "We've moved beyond the experimental with this project," Potter said, clomping the clear-finished hemp fiber tractor hood again. "We're into the implementation of these things. This is going to be a commercial product."

Potter explained that, although the CIC is a nonprofit and government-funded, the center's teams are allowed to be entrepreneurial. So when the tractor body's field-testing is done, they'll likely partner with a commercial vehicle-production company on the engine, transmission, electronics, and drive train—the moving parts, in other words. "Or," he said, "you can buy the hemp body from us and design your own vehicle."

Holey Gazoley, I can't wait to see *that* in the John Deere or Caterpillar showroom. A documentary called *Government Grown* mentions that International Harvester once made a combine that worked "with the tallest hemp stalks." Surely that blueprint is somewhere.

My hope is that the CIC and its partners will work the necessary features into the final product and release it on a commercial scale. That'd certainly be useful to our hemp pioneer Grant Dyck, whose harvesting equipment burst into flames (twice) in 2012. I hereby offer my online handle of Organic Cowboy as the tractor's model name in exchange for a reasonable franchising fee. Though The Hemp Reaper is pretty good, too.

I'm kind of expecting a call on one of these, based on Potter's industrial intelligence report. "We have automotive industry designers coming by almost on a weekly basis asking, 'When are these materials going to be ready?'" he told me.

To address industry interest in all of its work, Potter said the CIC is developing a project called FibreCITY, which is a replicable franchising system for anyone who wants to open a fiber-processing facility "appropriate to their regional cellulosic needs in Kentucky, Australia, or China." Could be you.

That morning's tour made me further optimistic that my descendants will have a breathable atmosphere without humanity returning to the Stone Age. The CIC is demonstrating a real-world mechanism by which a sustainable industrial cannabis infrastructure can reestablish itself.

There is a wrinkle to work out with the industrial and high-tech sides of hemp, though. Though each bend in the CIC facility revealed another hemp marvel (I especially loved the load-bearing hemp wall used for its sound-insulating qualities by CIC engineers to keep their own loud compressor machinery from distracting them), some applications require what Potter calls "greater fiber consistency."

My tour guide that sunny subzero day was broaching a technical issue that, though hardly insurmountable, the North American hemp industry has hardly begun to address.

It goes like this: Yes, Potter says, hemp's most passionate advocates are correct that the plant's tensile strength is (or can be) greater than steel's. But farmers, especially in the young, seed-oil-centric North American fields, are not yet growing for fiber that displays this quality. It's possible that the perfect industrial cultivars don't yet exist.

Thus, Potter told me, "We're starting to work with farmers to breed the kind of consistent, strong bast fibers that are absolutely essentially for sophisticated technological and industrial applications." The whole conversation left me wondering, *Any grad students out there looking for a dissertation project?*

"We even have to examine the soil nutritive regime, which can damage the fibers," Potter said. He laughed, then added in a conspiratorial tone, "Today farmers beat the crap out of the fiber during harvest. And once we do come up with ideal cultivars for our applications, we have to be able to replicate the breeding processes."

Okay then, I thought. *Let's get to work.* I felt safe thinking that because those blessed Canadians have a hemp GMO ban prophylactically in place. Seriously, it's a cool country. Even though the rumors that Fox News is banned there aren't true. I didn't see it on anywhere, though.

CHAPTER SIX

Fill 'Er Up with Hemp

*I*t was midday on February 10, 2013, and I was *very smoothly* cruising south out of Denver, Colorado, in a hemp-powered limo. A sleek cream-colored 1979 Mercedes 300D, in fact, purportedly originally owned by Ferdinand Marcos.

"Plenty of space," the driver told me when I asked for the keys to the trunk before the drive. "Imelda's shoes aren't in there anymore."

Hemp oil was the fuel, but it's not the kind of thing that I, the pampered passenger, would notice if I weren't a cannabis journalist. The vehicle was equipped with a proprietary shock system that results in the sort of sensory experience I normally associate with water beds. The giant backseat (more of a backroom) sofa—indeed, all the seats—was covered in sheepskin. There was room enough for me to do my morning yoga back there.

Regardless of the unbelievably comfortable ride's lineage, Bill Althouse, the chauffeur, was trying to demonstrate something on this enjoyable winter outing to Colorado Springs. What the longtime sustainability consultant and renewable energy engineer was showing is that in 2013, a plant cultivated by humans for eight millennia can replace petroleum.

2013 is a year during which we same humans will no doubt surpass 2010's consumption of 37.7 billion barrels of oil—that's 87 million barrels a day, or a million barrels a second.[28]

It wasn't proving my least eventful road trip ever. We noticed before we hit Castle Rock that we'd slightly miscalculated the distance for our planned spine-o'-the-Rockies drive this day. But, with a few roadside top-offs, we pulled it off on fumes thanks to the excellent twenty-two MPG we got on hemp biodiesel running through a big old engine engineered for dictator spinal comfort, not for efficiency.

We wound up using seven gallons of hemp oil for our ride. The U.S. government says that petro diesel spews 22.38 pounds of CO_2 into the atmosphere per gallon. That means, since the statistics folks at the Energy Information Agency also say biodiesel releases 78 percent less carbon, that in this short road trip we prevented 122 pounds of carbon from, ya know, clouding the future of our species. And we didn't give a penny to ExxonMobil.

For reasons of scale alone, seed oil might not immediately prove the ideal part of the plant for exploiting hemp's energy potential, according to Althouse. Which is to say it'd take an awful lot of hemp acreage to prove cost-competitive at the pump. Sure beats fossil diesel, though, in odor alone. The exhaust smelled organic. Like all vegetable-oil-based fuel, it made one hungry.[29] Ran quieter and more efficiently, too, Althouse said.

The fuel itself, I earlier noticed as I watched the sixty-one-year-old Althouse pour it in from a heavy plastic container, was the fecund green of a lily pad, and it wasn't easy to acquire or process. Althouse managed to get his hands on some five-gallon "cubies" of food-grade Canadian hemp oil, which was then processed into biodiesel by a local Colorado outfit called ClearEcos.

Back in Imelda's limo, Althouse said that he'd like to see cannabis's fiber harnessed for energy rather than the seed oil we were using. At the very least he'd like to see oil cultivars specially developed for fuel, per Potter of the CIC's sense that a hemp application only works in the big leagues if the right varieties are utilized.

"This Finola [cultivar] we're driving on produces an ideal omega ratio," he said. "It's food, not fuel."

Here's why this matters: ClearEcos's Kurt Lange said that the very lignan and other nutritive components that make hemp seed oil a genuine superfood make it difficult to process to the viscosity needed for federally approved biofuel.

"We increased the process time to deal with the additional fatty acid chains," he said of my limo ride's tankful of hemp oil. It's not ready for prime time at the corner gas station, in other words, despite the carbon we kept from emitting on February 10 even with this prototype form of cannabis fuel. It'd probably price out north of ten dollars a gallon—until American farmers get those two million acres planted.

Lange's piece of behind-the-curtain processing information explains why at no time does my Vaporware Sensor sound more piercingly than when I venture into the energy sector. Probably because it's so vital to our Netflix, texting, and Cancun vacations (not to mention our food supply) that we resolve our little fossil/nuclear problem, we subject ourselves to a lot of premature excitement about Next Great Energy Supplies. Most of these, so far, have turned out to, at best, require more time and infrastructural investment than excites our insufficiently climate-concerned representatives.

Not that there's anything inherently wrong with huge investment in important endeavors. It took a massive investment to create,

HEMP PIONEERS

Bill Althouse, Engineer,
Hemp Advocate, Limo Driver

Everyone's an individual, but in the hemp world everyone's really an individual. A character. And thank heaven for that. If my over-educated hemp-oil-powered limo driver wanted to rant, like the limo driver in *This Is Spinal Tap* feels compelled to (and in fact as seems to be part of the baseline training in limo driver college), I was going to let him. I mean, c'mon, the guy was giving me the literal road-test proof of the value of hemp. This was a breakup letter to petroleum.

So here is the sixty-one-year-old Althouse's reply, from *way up* in the driver's seat of Ferdinand Marcos's limo (practically in the next county—I had to shout to be heard from the sheepskin coach where I was chillin'), when I asked him, "How does it feel to drive via hemp power?"

"Awesome. Absolutely awesome, of course. But I'd like it to develop into some other part of the plant rather than just the seed, or the seed variety we're driving on. Because this seed oil is one of those super-food deals. I think the stalk might be the first place we go for energy. But what I'm really saying is I'd like to see unfettered research taking place. For instance, here"—he pointed out the window as we passed the National Renewable Energy Laboratory's headquarters outside

Golden, Colorado. "And at Colorado State University—ninety minutes from here—there are brilliant people who have spent their lives understanding the biochemistry of plants better than anyone else in the world. They need to be working on hemp yesterday."

nearly overnight, a wartime economy that could beat Japan after Pearl Harbor. One day General Motors made cars, the next it made tanks. Indeed I keep waiting for President Obama to keep his campaign promise to put America back to work building a sustainable energy grid infrastructure. Make no mistake: Climate change (the sum of its causes and offshoots) is a Pearl Harbor, minus the instant explosions so useful for mobilizing public opinion.

The really good news is that hemp energy needn't be hugely expensive. Althouse's message is that, until we develop the right cultivars and grow them en masse, it might not be hemp at the gas pump we see first (via seed), but hemp at the power plant (via fiber). Which is, from a climate stabilization perspective, actually far more important to our species' survival. Why? Because 40 percent of carbon emissions come from power plants, according to Daniel Becker of the Safe Climate Campaign.

So if not hemp biodiesel (in the short term, at least), how exactly are we to transition from climate-altering petroleum and coal without giving up Netflix and Cancun? To answer that, we're a-headin' to Kentucky.

CHAPTER SEVEN

A New Utility Paradigm—
The Distributed, Sustainable
Community Energy Grid

*A*n entrepreneur in the Bluegrass State is out to craft a win–win scenario whereby locally grown hemp allows struggling rural communities in former resource-extraction economies to transition to a sustainable economy (in this case from coal and monoculture tobacco).

The first thing to know about Roger Ford, CEO of Patriot Bioenergy, is that he's not wearing a tie-dye. In fact, from where I sit there's nothing quite like hearing the CEO of a Deep South energy company extol, in a mild regional accent, the virtues of ending the war on cannabis.

"Hemp has historical ties to Kentucky," Ford told me. Except for this current break, "It's been a major producer since colonial times, especially for shipping. George Washington grew it in the South. Today we have land affected by surface coal mining. We can implement the use of biomass on former mine sites for reclamation. Federal law has to change so we can ramp this thing up. The obvious question is, why did we ever stop?"

Ford is not alone in wondering. Most Kentuckians today, like most Americans, want to end the drug war. And Ford is doing more than wondering: Patriot Bioenergy is the first corporate member of the Kentucky Hemp Growers Cooperative Association trade group.

Make no mistake: Patriot's in it for bottom-line reasons, and the plan is to use hemp's fiber. "When farmers are growing hemp for the seed and long fiber, what you have left is the hurd—your woody materials," Ford told me, displaying an innate instinct for dual cropping.

Remember when we wondered what happens to the hurd that's sitting around the field once the value-added applications have moved on to processing? Ford wants it to be turned into electricity, in part via combustion techniques like the gasification technology we've touched on. In the Patriot model, this will happen not on individual farms, but in a community-based "integrated energy park" that also includes ethanol production from regional sources of agricultural feedstock like beets.

Ford doesn't see a 100 percent hemp-powered Dixie, at least not right away. In fact, he's not even turning his back on Kentucky's coal. "We're looking at feedstock blending with coal and natural gas in a multiple-generation scenario," he said. "With gasification of biomass product—hemp and industrial waste—as the goal so we get a combustible material. It generates steam in a clean-combusting boiler, which produces electricity. These things are commercially viable on a large duplicated scale."

Despite hanging on to the energy c-word (Ford believes the addition of biomass will clean up coal emissions), what's appealing about the aptly named Patriot business plan, dressed as it is in a business suit with apple pie crumbs on it, is this: Ford is blueprinting a method by which communities can at once take control of their regional energy production, put themselves back to work, and heal the soil. Hemp can help farmland in a number of ways, but one of the most important is simple soil stabilization thanks to its long, fast-growing taproots. This phytoremediation, the process of using plants to repair polluted terrain, is in fact a key part of the hemp calculus in the climate change era. Plus, a decentralized regional energy model is more secure than a single mega grid that could all go down at one time.

But how will it work? "We finance and build plants in rural areas—producing to scale as needed locally," Ford explained. "And in the process create rural jobs. That's our model: rural economic development through sustainable energy production."

In starting the company, Ford studied how the petroleum market developed in the late 1800s, and was surprised to find small refineries all over the country.

"Historically, John D. Rockefeller and Standard Oil focused on facilities that produced a few thousand barrels a day," he said. "We're replicating those models. We're not talking about fifty-million-gallon-a-day facilities [that cost] hundreds of millions of dollars to build. We're mapping that strategy onto biomass."

The Kentucky legislature overwhelmingly passed a hemp cultivation ordinance on March 26, 2013, which the governor allowed to become law without signing. The Bluegrass State's key politicos are pretty much all on board, starting with its chief farmer. "We could be the Silicon Valley of industrial hemp manufacturing right here in Kentucky," Agriculture Commissioner James Comer told the *Kentucky Enquirer* in January 2013. (At the end of all these kinds of statements, you can add, "as soon as the federal drug war ends.")

Describing Patriot's ambitious utility-localizing strategy as "in the planning stages" (understandably, given hemp's dubious legal status for the moment), Ford said that he's got locations scouted from Kentucky to Mississippi. Patriot has signed a letter of intent with local government in tiny Whitley County, Kentucky, he added, for the first of these regional energy parks. This, incidentally, is the setting for the Grateful Dead's ode to the hard life of the coal miner, "Cumberland Blues."

"That facility will at first produce four million gallons of ethanol per year from sugar beet feedstock until industrial hemp comes online, for growing in marginal post-mining land," Ford said. "And it's replicable in other communities." This mention of marginal land emphasizes another quality of hemp: its ability to provide a viable harvest in areas other crops can't. Hemp has even been discussed for radioactivity mitigation around Japan's Fukushima nuclear plant disaster.[30]

I asked how much each energy park would cost. I'm glad I did. I love it when people sniff at sums that seem unfathomably large to me. "Oh, less than seventy-five million to build," Ford said. "We want to make this viable economically."

Vaporware alarm sounding faintly even though the company already has some beet acreage planted, I asked if the privately held Patriot, based in Bowling Green, had that kind of dough. Ford sounded confident. "We're having conversations with venture capital folks, and working with local and state government. They know this plan will increase the local tax revenue base, which is getting hammered in this economy, especially given the coal situation."

At this point in our interview, I wondered aloud how a onetime member of his college campus Republicans club came to hemp as a major feedstock player for his regional energy plan?

"It's proven that we Kentuckians can produce a [hemp crop of] quality second to none in the world," Ford told me with folksy pride. "I'm comfortable working with anyone on this. Democrats, Republicans, hemp activists."

Patriot Bioenergy does have history on its side. Kentucky really was the hemp world leader before federal prohibition kicked in after 1937. In 1838, "There were 18 rope and bagging factories in Lexington that employed 1,000 workers."[31]

Our limo-driving engineer Althouse thinks Kentucky's on the right track. "It works," he said of gasification. "You get very efficient energy out from the biomass you put in. If it's done right."

He should know. Far more than just a chauffeur, Althouse worked on a federal-funded report that advocated bringing in a biomass project to the Santa Fe, New Mexico, area in 2006. The wood waste energy in that plan would come care of gasification furnaces of a technology similar to (but, Althouse says, with thirty times the energy output of) commercial models manufactured by a German company called Herlt.

Page 15 of the New Mexico report has a lucid diagram that explains how the boiler at the heart of biomass combustion technology works. Although to a right-brainer like me it looks like Hawkeye's mid-tent $M*A*S*H$ gin still. Or something Wile E. Coyote would build when trying the laboratory approach to roadrunner hunting.

But you get energy from biomass with negligible carbon emissions. The plan was so doable (at less than twenty-eight million dollars per facility) that Althouse is convinced the Land of Enchantment was at the cusp of building a biomass infrastructure that today would be a perfect fit for hemp.

"Then fracking happened," he said. "And the project fell apart. We were almost there. The way the world works is *I'm green unless it costs me something.*" Althouse's take echoed Hemp Oil Canada's Crew, who said that the reason Canada hasn't been exploiting hemp fiber energy is "gas is still too cheap."

Althouse told me emphatically that the hemp/biomass energy speculator has to answer some comprehensive questions before entering the sector: "What's your plant start-up cost, for one?" he asked. "For, say, a small, municipal-sized, potentially multiuse community biomass plant" like Patriot's planning. "It's called a value chain analysis, and it has to be rigorous."

But Althouse remains a believer in gasification technology. "If it's a closed-loop dedicated energy facility, you're emitting almost no carbon."

He added that, just as Canada's Cramer was worried most about building regulations, he also saw a major bureaucratic hurdle to leap before implementation of a new energy paradigm. "If the emissions regulatory framework can be hammered out domestically, that'll be a huge step forward," he said. "They did it in Europe, but it could prove more difficult here, where in most states utilities control the system."

Here Althouse rattled off a litany of cards in a stacked deck against independent energy aspirants, before adding, "In Germany the citizens pay a buy-in tariff, and control their energy."

And therein, for me, lies the rub.

Who's not rooting for out-of-work coal miners out to restore their backyard and mitigate climate change while producing their own energy? The whole time I hung out with Althouse, who carries a colorful bio beyond his engineering pedigree that includes seven years in the South Pacific doing energy consulting with the government of Palao, he never let me forget this point: A biomass energy transition model loses its appeal if it just creates a new utility monopoly, albeit a cleaner one.

Indeed, when I checked out the appendix in that New Mexico biomass energy report on which Althouse worked, I saw that "community ownership" and local "energy trusts" are considered integral to the plan's viability. Whereas Kentucky's Ford, Althouse reminded me, is citing Standard Oil.

Energy independence, of course, is the Holy Grail of both our domestic and foreign policies. And after seven years of researching it (and trying to live it on my own solar-powered, goat-dominated ranch), I don't pretend to have all the answers or even all the questions. I'm just trying to fire up this debate and figure out what role all that hemp hurd we're going to have lying around by the thousands of tons in a few years might play. Wasting it is not an option.

The two words I promised Althouse I'd attach to that search are "community empowering." Really this was my only payment for the limo ride other than breakfast.

"Don't forget the farmer!" he shouted as I started my own vegetable-oil-powered rig in the street in front of his Denver house. My exhaust, as usual, smelled like Kung Pao Chicken.

I hardly had a chance to forget. As I was about to send this book to my publisher, the *New York Times* reported on August 17, 2013, that the U.S. Army, frustrated by fuel hauling hassles in danger zones, is embarking on an eight-million-dollar gasification test program, based on a furnace the size of a shower stall, called the FastOx. A community-sized model! Huzzah! This might be the military's greatest gift to society since the Internet.

But really, is distributed, locally owned and grown energy a viable plan in the real world? Is there anyplace where it's already happening? The answer to those questions is a resounding yes. It's happening in Germany and a few other spots. Not on a wide scale with hemp yet, but with other biomass from local crops.

To a fan of local and renewable energy and an opponent of resource wars, the website of the town of Feldheim, in the eastern German state of Brandenburg, is a thrilling read. Or as thrilling a read as a technical energy production site can be. Right there in the first paragraph, it tells us that the town's five-hundred-kilowatt biomass plant is "owned by the local agriculture collective." Althouse's primary problem solved.

Though it was already famous for turning 30 percent unemployment into 0 percent by putting everyone to work at the community energy park, I first heard about Feldheim from a 2011 Associated Press article.[32] It pointed out that the energy and the jobs all came when consultants suggested stuffing the local agricultural waste into a gasification and methane (natural gas) harnessing plant.

"It is possible to live completely from renewable energy," German transport minister Peter Ramsauer glowed in the article.

None of it means anything to a guy like Althouse without that "owned by the local agricultural collective" sentence. I'm not arguing. Even though, all other things being equal, I tend to wear a free-market hat, when it comes to energy, things are not equal. We've got our children's air and water at stake, we've got a silly utility system and an antiquated grid. I am more than slightly open to alternatives like the Feldheim miracle in my own community.

The project at once put one of Brandenburg's most depressed areas back to work and made it energy-independent. And Feldheim is not the only German town living the dream. I'm keenly interested to see if the European experience is mappable to the United States.

Is regionally based renewable energy modular? Will distributed biomass plants make small-town America (and big cities for that matter) energy-independent? I'm really asking you, the folks reading this, if you have the engineering know-how or venture capital to make it happen.

Now, I can tell you after spending half my life in extremely rural communities that organizing farmers to unite, even in their mutual best interest, is a bit like asking snowflakes to canvas for Democrats in Houston. But when the future of the species is at stake, maybe we'll see some overdue rallying.

Gasification, of course, isn't the only way to turn biomass into power. But it exists in the real world and has a relatively low start-up cost compared with, say, the $350 million that BP is investing just to upgrade a Brazilian ethanol plant.

That kind of alcohol making for power and driving sounds terrific, but besides the fantastically high facility costs, ethanol requires sometimes unsustainable catalysts, and depends, in the case of Brazil, on forest-depleting sugarcane monoculture. As an industrial-scale technology, ethanol might very well mature. Hemp would work in such a scenario. I chose to focus on biomass combustion herein to show one solution whereby farmers and communities are already becoming energy-independent via their farm waste. As we've established, anything's better than torching it in the field.

There are still other biomass technologies out there. Methane can be harvested from livestock and ag waste. That could figure into, say, a hog-and-hemp-farming community's formula. You can even feed the hogs the hemp seed cake harvest, which consultant Anndrea Hermann does on her Canadian farm.

Althouse said he's also keeping his eye on a cellulosic process now being developed in British Columbia known as lignol. This, he said, takes not just your hurdy fiber but the whole plant and "leaves you with a pure lignin[33] that can be used for paper pulp, fine clothing fiber, or a great spray insulation. GE developed it during the first oil crisis, in 1973, and it's easy to re-create the process in the lab." Sharp guy, that Althouse. I look for that in a chauffeur.

Then there's the good ol'-fashioned hemp biodiesel that got Althouse and me around Colorado in such cushy fashion. It's quite feasible that just as hemp cultivars are being developed for consistent high-tech industrial applications, so might there be a seed cultivar ideal for fuel—one that won't take away from food.

Oh, how I wish Warren Buffett would stimulate this market by converting his BNSF trains to run on American-grown renewable hemp fuel instead of natural gas. A Grand Canyon tourist train already runs its diesel engine trains on vegetable oil. The United States almost certainly has the acreage for all the industrial cannabis applications that could ever come to mind and then to market.

Know what else she has? Thousands of small farmers desperate for a new cash crop that'll grow without much water as climate change continues to hammer the heartland.

In the end, all of these promising initial hemp applications might prove to be "bridge" technologies: part of a nation's or community's transition from petroleum. Personally, I like technologies that don't have to burn anything. *Harness* is the word that comes to mind for me, as in "sun," "plants," and "wind."

It's my hope that this project brings thinkers together. We've planted a bunch of interconnected preliminary thought seeds here in a soil that I hope will grow like a Manitoba hemp field (minus the combine fires). There is a template for biomass energy in place, and no one knows better how successful it can be than the residents of Feldheim.

Whatever the method and source, U.S. federal renewable-fuel standards mandate that our nation produce thirty-six billion gallons of biofuel per year by 2022. We'd be wise to make sure the *Cannabis sativa* plant is a major part of that, for the good of the economy and the atmosphere. Which is another way of saying, "So we don't die out and so we can stop killing one another over black dinosaur jelly."

CHAPTER EIGHT

Don't Just Legalize It— Subsidize It

*B*y the time I clued in to communities like Feldheim, I was pretty excited. The whole picture was in place: seed oil, construction material, energy. Hemp hemp hooray.

I remember gleefully perusing my gasification file at the tail end of my Canadian hemp research, just after informing the U.S. Customs man that I carried no hemp home with me "other than my breakfast, lunch, pants,[34] shirt, hat, and soap." *Game on for this industry*, I thought.

Then, back in comparatively toasty New Mexico, I spoke to some European consultants who kind of brought me back to Earth. "See," they explained, "there's this thing called the real-world economy."

Even with all of hemp's exciting species saving and climate change mitigation, a German hemp expert named Michael Carus told me I shouldn't expect a profitable American market to leap magically into existence the moment domestic cultivation ramps up.

In fact, it might need help to ramp up, he said. Income from domestic hemp cultivation for fiber, especially, wouldn't be competitive enough on the free market to incentivize American farmers to grow the millions of acres we need for our dual-cropping, humanity-saving plan. He said China, to give one example, grows textile fiber cheaper than America would.[35]

What a downer (sorry, realist) that Carus was! Of all the hemp experts I interviewed, he was the one who seemed patently uninspired by the American hemp sector coming online. I got the impression that most of our hour-long conversation was, for him, an exercise in reticence and caution, no doubt well learned. Indeed, the European market, though steady and robust at a hundred million dollars annually, wasn't projected to show Canadian-level growth in 2013, and there was even talk of a seed shortage.

Still, I sometimes think these Europeans willingly fail to figure American exuberance into their economic formulae. That's our real fuel. That, hemp oil, and love are pretty much all I run on. They and indeed all economists can call it Σ or ® or something and consider it a constant that makes any venture ten times more likely to work. Some folks might think I'm kidding. Actually, persistence and optimism, basically America's two required traits, are always listed among the most vital qualities cited by today's successful entrepreneurs in your finer airline magazines.

Whatever the reason, I just couldn't get Carus psyched about American hemp prospects. That is, until the very end of the interview. It was only when I asked him, "What if official U.S. policy incorporated the true economic value of the cannabis plant, including soil remediation from hemp's, ya know, famously deep taproots?" that Carus finally burst into an almost New World exuberance.

"If you can convince Obama to implement European-style subsidies, the U.S. market'll be okay," he blurted, albeit with a sort of deep chuckle. Then he added, quite seriously, "Like Europe, American agriculture is guided by government incentives. If your government decides it wants to encourage hemp, well . . ." At this Carus's face melted back into mirth, and he emitted a sort of "ho ho ho!" from beneath what was actually something of a Santa Claus beard. For some reason a lot of the folks, worldwide, who work in hemp look like Dori or Nori.

The chuckling no doubt had to do with the four hundred dollars per acre with which the European Union subsidizes its hemp farmers. He was wishing us luck.

The British hemp expert John Hobson we earlier met, who advises European hemp farmers on agricultural nuance, emphatically concurred with Carus's assessment. Especially when I told him that our president has vowed to cut U.S. greenhouse gas emissions 17 percent by 2020.

"It's rare enough that hemp's a soil builder," he explained, "but it requires no spraying of pesticides, herbicides, or fungicides. If those qualities are worked into an agricultural subsidy formula, it suddenly looks much stronger."

And that's before you get into carbon sequestration. Hobson's company website says its version of hempcrete "locks-up approximately 110 kilograms of carbon dioxide per cubic meter of wall."[36] Greg Flavall said that's comparable to how his North Carolina house performs in its role of mitigating climate change.

So how do we make the numbers add up to profit for farmers on the home front? There's always a hemp subsidy we could drop, hardly noticed, into the next FARRM Bill. At least to start, to help the industry get on its feet across the United States, until those two million acres begin to literally seed our food, construction, and energy revolutions.

There's another way to encourage domestic hemp farming, too. One that consciously aims to return America to her family-farming roots. We are, incredibly, down to an embarrassing 1 percent of Americans farming now. It was 30 percent the last time hemp was legal.

Just as the famous Homesteading Act wound up feeding the nation, I've heard a new plan described in the form of a Digital Age Homesteading Act. This would provide industrial cannabis farmers in places like North Dakota, Vermont, and Kentucky with micro grants for land purchase and cultivar research. Funding can come from the two billion dollars President Obama committed to alternative fuel research in March 2013.

And on that encouraging note, we return to our hemp entrepreneurial journey. We'll stick for the moment with construction materials like hempcrete, since that's our first fiber killer app. Without question, Carus is correct that the expensive hassle of the drug war has things in something of a holding pattern. Unprocessed hemp doesn't travel or store particularly well or inexpensively. Which is to say imports of anything are rarely cheap.

But domestic hemp will win in a worldwide fiber marketplace that has a level playing field for raw materials, the Hemp Industries Association's Steve Levine believes. That's simply because it is "too amazing a construction [material] not to, not just for insulation, but for load-bearing block components, roofing, paneling, fiberboard, and flooring." And when high international shipping costs are eliminated, he said, that is a big step toward a level playing field.

Furthermore, said Vote Hemp's Eric Steenstra, domestic hemp will be profitable even during the high-end fiber processing learning curve the CIC's Simon Potter believes is necessary, because "fiber for industrial composites and construction can be successfully shipped in a more raw form than textiles."

That, in the end, is why builders like Flavall and Tim Callahan are having a go at it even as the drug war's final fires are being extinguished. They're confident that they'll find affordable hempcrete in tomorrow's Hemp Home Depot. They want to be the established industry leaders.

In addition to exuberance, the United States already has one huge practical advantage over much of the world: the factory infrastructure necessary to launch our fiber apps from season one. Want proof? Simon Potter thrust a sort of thin, strong, fibrous blanket at me during my CIC visit and then told me that the Canadian team that developed the "extremely promising hemp insulating mat material" had to (get this) *send it to Pennsylvania* for the actual fiber weaving. They just don't have the processing facilities in Manitoba.

For this reason, hemp advocate Norm Roulet, who is trying to organize the legalization and maturation of hemp in Ohio, believes that the Buckeye State's unparalleled farm-to-rail-to-factory network is ideal for the industry's rebirth. So ideal, he said, that Ohio should be the seat of a hemp commodities market. Seeming to confirm this is a 2006 USDA ethanol feasibility study that asserted, "The optimal location of a . . . processing facility is largely dependent on being in close proximity to its feedstock supply, regardless of which feedstock is being utilized."[37]

This is terrific news for a place like Ohio if its economic planners want vertical integration and local job creation to allow the state to again emerge as the major hemp player it was in the nineteenth century.

"We have more factory space in Cleveland than we know what to do with," Roulet told me. "Rail lines deliver right to it from nearby

HEMP PIONEERS

Don Wirtshafter, Founder and Owner, the Ohio Hempery

When Wirtshafter opened his legendary Hempery shop, (oil) press, and (publishing) press in 1991 in his third-floor Athens, Ohio, law office, most people thought he was a pioneer—an early adopter. But that's not exactly how he sees it, even if he did supply much of America's hemp oil through most of the 1990s.

First, there was local history, which he'd researched heavily. "The first cash crop of any kind in my area was Joshua Wyatt, in 1727," he told me. "He took twenty-seven bales of hemp up the Ohio River without a boat. He made rafts of it and sold it to the river boating industry for rope."

Industrial cannabis was so important to the Buckeye State by the nineteenth century, the sixty-two-year-old Wirtshafter said, that "Ohio State University's trustees allowed tuition be paid in hemp." In my own research, I found references to "broad expanses of well-shaped fields of hemp"[38] from the 1820s and 1830s that were inspected and praised by U.S. hemp agents and sent to local processors.

The second reason Wirtshafter doesn't totally grab at the pioneer mantle is because of the reason he came to hemp.

"I went to law school for environmental reasons, but my friends kept getting busted for pot," he told me. "When I returned home, I was doing some environmental work, but dealing with a lot of pot cases at a terrible time. You had Nancy Reagan, mandatory minimum sentencing—people going away for ten years for two joints—and the good-faith exception [which allows illegally obtained evidence to be admitted in court in some cases]—all the defenses that were taken away that made criminal law so gruesome in the 1980s. I saw my whole peer group getting taken down or at least getting very secretive about this plant. Finally the era started winding down, and I saw a way to unite the environmental and the criminal justice issues I faced in my small-town law office. I thought if we could wake the world up about hemp, if we could untie the knot around this plant, it would help both."

The Ohio Hempery sold your hemp oil (one of two U.S. sources in the 1990s), your hemp hats, and your hemp yarn, and "reminded heartland of its agricultural heritage," according to Wirtshafter, who still has an active law practice. "But," he said, "the real product was the cookbook." This was *The Hemp Seed Cookbook*, which Wirtshafter co-wrote with Carol Miller in November 1990. It sold ten thousand copies and is for now, sadly, out of print. "I'll get it online now that you've reminded me," Wirtshafter told me.

Wirtshafter said the cookbook "had the first description of how to shell the seed to make delicious hemp milk at home, with a colander and blender. The stuff that's five bucks a quart in your grocery store today. There were recipes for toasted hemp seed tamari salads, and there were all kinds of sprouted seed recipes." Wirtshafter did not pause when I asked for his favorite recipe in the book: "Hemp chocolate chip cookie," he said. "I realize now I ate differently then."

The Hemp Seed Cookbook, with a pound of hemp seed stapled to the cover, cost seven dollars in the Hempery, plus postage for shipped books that Wirtshafter said the Hempery "advertised heavily. You could call 1-800-BUY-HEMP. The plan was to spread the hemp message worldwide. The business grew quickly."

This went on from 1991 to 1998, when the DEA shut down the New Jersey mill the Hempery was using to sterilize the seeds—they actually had a schedule one permit to import the seeds.

"That was the year Canada's market opened up," he said. "So I wound up touring the north giving advice to our competitors." Fortunately, he said, the American market's launch is now "inevitable"—though he warned against measuring everything in terms of a coming billion-dollar industry. "It should be cultivated in everyone's backyard. We don't have to turn hemp over to big business."

Wirtshafter's other hemp-related exploits include driving in a hemp car that "did laps around the U.S." in the 1990s, including a California leg with longtime hemp activist Woody Harrelson. "The Hempery was making money," he said. "And we supported projects like that," as well as an abortive 1994 plan to plant a test crop at a USDA-recognized research facility in Imperial County, California (state authorities chopped the half acre down).

Looking at the coming hemp era, the pioneer (he is one) said, "Twenty years from now we'll be building our houses from hemp." Wirtshafter paused for a moment, before adding, "Again. Did you know that the [1883] Chicago World Exposition's buildings were finished with a hemp plaster called 'staff'?"

farms. With demand for industrial cannabis products so great, a historical hemp trade, and a perfect cultivation climate, let's put people back to work. The best part is that small farm communities and city neighborhoods—both in need of economic development—will be helping one another."

One Pennsylvania weaving plant is terrific. But does the United States really have sufficient infrastructure to support hemp? As with oil processor Shaun Crew's intent to parachute into the North Dakota seed oil market the moment it's legal to grow hemp Stateside, I think it's worth noting where the already successful hemp building companies are putting their money.

Our British friend Ian Pritchett, Lime Technology's VP, struck me as a savvy, unsentimental businessperson. His company wouldn't devote resources to a market if its "sums," as he put it so Britishly, didn't make the accountants smile. He went so far as to tell me, "Every new market is a battle." And Lime Technology is evidently so jazzed about the United States coming online that it's already started the Chicago-based subsidiary called American Lime Technology we earlier discussed.

This is a company from which you can today order hempcrete for your next building project. It'll be imported hemp, for now, of course, but what're you gonna do, beyond calling your congressperson and senators to bend their ears about the Digital Age Homesteading Act?[39]

CHAPTER NINE

Patriots Ponder Planting

*B*ack in the waiting American fields—the millions of ready-for-hemp acres in North Dakota, Kentucky, Ohio, Wisconsin, Oregon, California, and Colorado, to name a few states with either significant historic harvests or modern hemp-friendly laws—the inability to capitalize domestically on these market forces is intolerably stifling.

With the tide turning their way, some farmers aren't waiting. Coloradans explicitly legalized industrial cannabis farming in the same 2012 election that permitted adult social use of psychoactive cannabis. As many as two dozen farmers in the Rocky Mountain State planted hemp in the spring of 2013. One of them, fifth-generation Colorado rancher Michael Bowman, told me he's quite willing to be a test case, on the agricultural side and the legal side.

"We can eat it, wear it, and slather it on our bodies, but we can't grow it?" posited Bowman, whose *Aw shucks, I don't know better than anyone else, I'm just tryin' to do what's right* humility belies both his political savvy and his ranching know-how. "That's inexcusable. It's shameful. Do our federal drug squads really want to raid a longtime family rancher for growing the fiber the Declaration of Independence was drafted on?"

HEMP PIONEERS

Michael Bowman, National Hemp and Sustainability Lobbyist

"Check your email—I just shot you a photo of me with [U.S. agriculture secretary] Tom Vilsack," Colorado's Bowman told me when, in April 2013, I had asked, "C'mon, really? Hemp is going to be federally re-legalized in this session of Congress? It never gets more than a snicker in committee."

"At least in the House," he said.

He was right (and I still have that and other Beltway action photos he sent during what turned out to be the partly victorious whirlwind hemp effort in 2013). The tide had turned. Common sense had prevailed. Or maybe the money the Canadians were making had finally talked.

It's safe to say that Bowman (along with years of effort by other hemp activists) was a key figure in the lobbying effort that saw Colorado representative Jared Polis's groundbreaking hemp research bill sneak into the FARRM Bill, thus (if Congress finishes the job in 2014) ending one of the most counterproductive agricultural bans in human history.

A towering fifty-four-year-old farm boy intellectual who emerges from weeks of communication darkness to call me at weird hours from weird time zones after tracking down cabinet officials and key congressional "maybes" on hemp, Bowman laughs like a good ol' boy, never has a negative word for anyone, and gets things done by operating according to a sort of Zen-inspired seven-year patience plan.

"It started with a community center I worked on [he's from a small, conservative farming town of 2,354 called Wray] back in 1983," Bowman told me over doughnuts in Denver. "I saw that a project that takes six months to accomplish the goal isn't even enough of a challenge for me. Renewable energy was years in the wilderness. Hemp was years."

Bowman's graying around the fringes and not wearing a peace sign t-shirt, "always a plus on the Hill," he said. After he was instrumental in his home state's passage of the first substantive renewable energy requirements in the nation in 2004 (for which, as a resident of downwind New Mexico, I am very grateful), Bowman realized what a difference an individual can make in state politics.

So he thought he'd see if that was true on the federal level. He paid his own way to DC in 2013 and crashed with friends while pounding the legislative and executive branch hallways every day for two months to speak the truth about hemp. And what did President Obama say when he bent the POTUS's ear about hemp as a biofuel source during a 2012 Oval Office visit that he'd earned as a White House Champion of Change for his renewable energy work? "He listened respectfully," Bowman said.

"This is not a new crop," he told me, draining his coffee mug. "We're just late to the game in recognizing its value in the digital age." Indeed, at least thirty countries cultivate industrial cannabis today.

One of Bowman's key political skills is that he can speak eastern Colorado rancher-ese. No one accuses him of being a hippie. Raiding his family farm would end the federal war on hemp in about a week, just from right-wing outrage.

Could happen. He was explicitly threatened by a DEA agent on NPR on January 28, 2013.[40] Even longtime hemp activist Adam Eidinger, a man who's been handcuffed for planting hemp seeds at the Pentagon, said that farmers like Bowman, if they plant before the drug peace officially breaks out, are "literally betting the farm."

As with all bets, there's a payoff for the winner. Like the North Carolina hemp builders, what Bowman is trying to get a slice of is the drug peace dividend: the billions that transfer back to the economy when the drug war budget is redirected. Besides saving taxpayers tons of money during a federal debt crisis, the coming era allows hemp to take off as a profitable commodity. Bowman's no fool.

Harvard economist Jeffrey Miron sees $46.7 billion just in annual tax benefits when the drug war ends. He's talking about the legalization of all forms of the cannabis plant, but we already know the industrial side will be a significant part of the new cannabis economy. As Bowman frequently reminded me, the rest of the industrialized world has a considerable head start over the United States. Luckily, American academia is starting to educate a new generation of hemp farmers and entrepreneurs.

HEMP PIONEERS

Lynda Parker,
Hemp Advocate, Grandmother

I met the gray-haired, dignified Parker (and these qualities are important, as we'll see) at the August 1, 2013, official hemp flag hoisting above the Colorado statehouse in Denver. The matriarch of hemp in the Rocky Mountain State was beaming here in the city in which she lives and had worked for decades as a yellow page directory sales rep. "Farmers are planting, I consider this achieving the goal," she told me.

What I discovered from the love Parker was being shown by the comparatively latter-day hemp activists that day at the statehouse was that Colorado's farmers and entrepreneurs are leading the United States into the billion-dollar world cannabis industry in large part because of the preparatory work done by this single human being.

It all happened because when Parker retired from the phone book sales job in 2005, she took a year off to decide what she wanted to do with her life. She only knew that "environmental values" comprised her criteria.

"I remember where I was when it came to me clear as day," Parker told me as state police unfolded and raised the flag made from the same material that Betsy Ross used for the first American flag. "It was hemp in neon letters. Hemp was the biggest difference I could make for the planet as an individual."

The now sixty-three-year-old grandmother had no previous lobbying experience of any kind. And yet if this industry takes off as predicted (remember, Canada can't plant new hemp acreage fast enough to keep up with demand), there will be buildings named after her one day. That's because unlike Kentucky and Ohio, Colorado doesn't have a traditional hemp industry. "This is about rescuing wheat and corn farmers who are losing their soil due to monoculture and climate change," she told me. "About a modern cash crop in an expanding area for our agriculture industry."

Parker's backstory—and Colorado's hemp head start over the rest of the United States—reads like a Choose Your Own Adventure novel. If a series of crucial happy accidents hadn't happened, hemp cultivation probably wouldn't be taking off in the state full-bore in 2014. To name one, ten years before her post-retirement hemp catharsis, in 1996, Parker had taken a political science course at the University of Colorado at Denver. As part of a "follow a bill's life cycle" project, she happened to be assigned to cover the nation's first modern hemp legalization bill (sponsored by Colorado state senator Lloyd Casey, it passed the state house but failed in the senate when the DEA complained about it).

"Had I not taken that course, I would not be talking to you today, and this hemp flag might not be flying above the capital," she said. In other words, hemp would probably not be legal in Colorado.

From the class she learned how the legislative branch of government works. She used that knowledge a decade later, in 2006, when she spearheaded her first hemp initiative. "The first thing I did was call my friend [Colorado state representative] Suzanne Williams [D–South Aurora]. I gave her my poli sci class final paper and asked, 'Can we revisit this issue?' She said, 'I think we should.' Suzanne became my champion, introduced me around, put me in touch with not just elected officials but the amazing and effective sustainability activist Mike Bowman. We pounded the hallways seemingly in vain for years. It was a lonely time."

Mark these works carefully, ye who hath given up on representative democracy: After those few years of blank stares and giggles, Parker changed the hemp laws in a big state, in a time of supposed corporate control of government, nearly alone. She had no political experience. Her secrets? "I love what I do, I dress conservatively, and I don't give up."

Not that she didn't consider throwing in the towel—more than once. "Oh, I told friends several times, 'This is hopeless, and nothing's ever going to move through the legislature on hemp.' But Suzanne, Mike, and I kept prodding and poking around to see where we could get an opening."

An early opening came from north of the border. "The Canadian consulate's agriculture people in Denver were very supportive," Parker told me. "By allowing us to use their conference room for meetings, they legitimized us. And they provided us with a huge amount of information about the hemp industry, which was really taking off for them. The RCMP conferenced with Colorado law enforcement, telling them they had no problems with their industry. Zero. That got our law enforcement on board very early, which has proven very helpful."

Still, Parker spent most of her time during those first years answering *Can I smoke my drapes?* jokes from legislators and Rotarians.

"It was frustrating," she said. "But we'd get little bursts of momentum, and by 2010 we were having serious conversations. I realized we were seeing a shift in the consciousness."

That year Colorado passed a resolution in support of hemp legalization that went out to the White House and Attorney General Eric Holder. "It was toothless, of course," she said of that first victory. "But it stated the real issues farmers are facing—water shortages, debt, and the truth about hemp as a soil restorer and cash crop."

Was Parker's age and buttoned-down sales experience an asset? "I don't think there's any question," she said, her hair prim and her sweater buttoned even this day. "I am a mainstream face for hemp. It doesn't get any more mainstream than a gray-haired lady who sells yellow pages advertising. No one was threatened by me."

Hemp's first actual legislative victory in Colorado came in 2012. Another Colorado hemp advocate, Jason Lauve, with amazing alacrity, helped write a hemp phytoremediation (soil restoration) bill, HB12-1099. "Representative Wes McKinley called me into his office," Lauve remembered. McKinley is a cowboy poet from a rural district. "He was there with two attorneys, and he said, 'We're gonna write a hemp bill today.' In forty-five minutes we had it drafted."

Lauve reached out to Parker to help the bill gain traction, knowing she had good contacts and unwavering intensity. The forty-two-year-old Lauve, who runs a hemp industry building clearinghouse called Team Hemp (one of its projects is a hemp house), along with advocate Dr. Erik Hunter and Parker, testified in support of the phytoremediation bill.

"It moved so fast with so much support—that's what was so rewarding," Lauve said. The phytoremediation bill became law when Governor John Hickenlooper signed it on June 4, 2012. Parker looked at it from the perspective of years of work. "We had educated the legislature. They were ready."

Then came another huge unexpected boost, a chapter in the Colorado Hemp Choose Your Own Adventure.

"Years ago," Parker explained. "I had told Brian Vicente [one of the leaders of the successful Amendment 64 voter initiative that legalized all forms of cannabis in Colorado in November 2012] that I didn't want to be active in the psychoactive side, since legislators were just starting to understand hemp. And yet he still included hemp in that initiative. I bow down to him in thanks for that whenever I see him."

To codify the will of the people on that count, the legislature, again with near unanimity (one senator thought the bill too restrictive), passed a bill (Senate Bill 13-241, signed into law on May 28, 2013) that will allow commercial cultivation of hemp in Colorado regardless of federal law. Farmers will have to pay for a state permit, provide their field's GPS coordinates, and verify the crop's low THC levels, according to Bowman.

That law created a hemp advisory committee whose members plan to have hemp cultivation guideline recommendations for the state agriculture commissioner in place in time for the 2014 cultivation season, said Bowman, who's a committee member. Given that those roughly two dozen farmers planted in 2013, before the state regulations were even implemented, it's anyone's guess how many Colorado farmers will give hemp a try in 2014 when it's fully kosher on the state level in any amount. Thousands, hopefully.

In addition to unlimited commercial cultivation for plants under 0.3 percent THC, Senate Bill 13-241 also allows research crops of up to ten acres for plants that for now might have higher THC, in order to develop seed stock with different traits. "We had thirty-eight sponsors for that bill," Lauve recalled. "It passed through every committee unanimously. That was it. Colorado is totally behind hemp."

"Another big part of why the state moved so fast is Colorado farmers said we're doing it," Parker told me. "They don't need DEA approval and they're not waiting for it." As for federal legalization of hemp, she added, "The momentum is utterly unstoppable."

So what's the message for activists in any cause? Parker had so many suggestions to tick off, it was as though she had waited her whole life for the question. "There has to be that level of maturity," she began. "Include the people you think will resist. Most of the time your supposed enemies just don't understand. Always take the high road, no matter how weird it gets—and it gets weird in politics. And most of all, try to have fun along the way. Looking back on it, I can truly say it's been totally fun."

Ya know, nearly single-handedly laying the groundwork for what looks to be a billion-dollar industry for your state's farmers. Not a bad thing to check off one's bucket list.

CHAPTER TEN

Hempucation
Immersion Course

*A*nndrea Hermann, strawberry-blond hair flayed across the back of her coveralls, was trying to help me dig my full-sized four-wheel-drive rental truck out of another snowbank, this one behind her former Mennonite farmhouse. Based on the fact that I could see only a frozen pancake in all directions, it seemed to me that I was in the dictionary definition of the middle of nowhere—somewhere on the Canadian Prairie. In her directions, Hermann had described the location as a suburb of Winnipeg.

The minus-seventeen cold at the moment—calculated before healthy breeze—was a physical presence on Hermann's 120 acres. An entity to which I could and did speak. And we were out in it for longer than humans are rated for.

I'm relating what any Canadian will tell you is just another Far North Thursday to get to the reason I was stuck: Both one of the world's most prominent hemp industry authorities and I had wanted to get ten feet closer to the former's barn to unload the truck. The surrounding snowdrifts, if not Himalayan, had seemed imposing to navigate on foot in such conditions. It was a lazy move, and we were paying for it. Good thing I had about a month's supply of Colleen Dyck's GORP bars in the cab.

In our defense, Hermann had tossed perhaps thirty white feedbags, at fifty pounds each, into the Ford's truck bed when we'd visited the Hemp Oil Canada facility. I'm pretty sure no previous Payless renter had done that. Not with this cargo: The bags contained pure hemp seed cake, the protein-rich by-product of hemp oil pressing. Upward of three-quarters of a ton of it. It was feed for Hermann's pigs.

"This and compost is all I feed 'em in winter," she told me just before I drove into the crystalline snowdrifts (where I remained for several hours) on February 21, 2013. "In the summertime sometimes they graze the hemp stalks in the field."

I watched the morning feeding: These were healthy-looking pink-and-brown pigs, energetically charging me either for pets or for more hemp. One of them was pregnant.

Hermann, in other words, doesn't just work to promote hemp. She lives it. Thus these are heady times for the thirty-six-year-old, not just because her consulting business line is ringing off the hook, but because she's still, after fifteen years in the hemp business, ticked off that hemp isn't legal to help the economy in her native Missouri.

"It's important that we win this, and we're nearly there," she told me buoyantly (or perhaps just shiveringly) back in the powder-swirling cryogenics lab of her farm. Billows of steam followed the words out of her scarf and around her swine-muddy coveralls.

Sometimes when she was animated, which appeared to be roughly the hours between 7 a.m. and 11 p.m., I noticed that Hermann accelerated from an erudite, deliberate, Canadianer-than-thou pace to the more rapid-fire Ozarks syntax of her upbringing. At one point I had to count the number of negatives in a sentence to see how she really felt about a particular Kentucky hemp legalization bill.

This is a Get 'Er Done kind of woman, one whose frankly barely habitable winter ecosystem kept making me think, *You have the skill set of the successful Fortune 500 CEO; you must really care about the hemp plant's reemergence.*

She does. Her terrifyingly drafty house is quite literally a hemp museum. There are hemp bales outside, 120-year-old hemp newspaper ads on her living room table, and hemp soaps in the bathroom. This is a woman who got into hemp as an undergraduate because her adviser asked her, "What needs to be changed so badly it makes you angry?"

A decade and a half and two degrees after throwing up her hands at the U.S. drug war, she's still at it, having never wavered, changed careers, or abandoned the Far North. And now she's consciously about to experience victory in her life's biggest battle. The end of the war on hemp, for Hermann, is like De Niro's LaMotta finally claiming the middleweight title in *Raging Bull.*

On the day of my visit, she chopped and chucked firewood into her woodstove with the body language of a slugger swatting batting practice, and then cooked her lunch on it, including those scientifically proven super-healthy hemp-fed chicken eggs that I'm rather craving three of at the moment. She did all this, as we've seen, while answering calls from her Canadian, African, and European hemp consulting clients about whether certain harvest techniques risked unwanted hemp seed sprouting prior to processor delivery.[41]

Not long after the pig feeding, as I wriggled like an outsmarted Houdini out of the coveralls Hermann had loaned me for the attempted truck extraction, she was on the horn with a college administrator in Corvallis, Oregon. They were discussing the landmark industrial hemp course that Hermann was co-leading at Oregon State University in the spring term of 2013.

This was a twenty-three-hour, 400-level class in the college of forestry at an American state-funded university. The course description was: "Introduction to the botany, biology and agronomy of the hemp plant, and the origins, historical contexts and implications of contemporary legal and social issues surrounding its use for food, fiber, and building products."

The first iteration of the course had thirty-six students. "The class was really engaged," Hermann told me a few months after my first visit. "And they were excited about creating their own bit of hemp history. Some students knew a lot about the cannabis plant, some none at all. But each brought his or her own expertise into what the course is offering."

Hermann said it was particularly gratifying to have total support from the university. "For fall term we're increasing the student number to one hundred," she said.

And so the hemp knowledge base is being rebuilt—OSU is like that Irish monastery that saved all the Greek and Latin classics. It actually makes sense that Oregon State has relaunched hemp into mainstream academic legitimacy, incidentally, as it was Beaver researchers who did some of the last pre-drug-war domestic hemp cultivar research in the 1930s.[42]

I personally didn't see how Hermann had time to prep for the course, since she's one of the people the Canadian government hires to sample hemp crops to make sure the THC levels are sufficiently negligible. Hermann sampled twenty-one thousand acres over the 2012 season, running from farms all over the province to a lab near her home, all while fielding that stream of hemp client calls.

The Hemp Industries Association's Steve Levine calls the Oregon State course "a big deal," since the plant is as kosher as basil locally but, obviously, not yet in DC. It certainly says a lot about American academic hearts and minds. Plus I have little doubt, after how much I learned from Hermann, that it was a valuable educational experience.

Levine sees the OSU course, Bowman's farmer activism, and national lobbying efforts as all part of the same final push to a domestic hemp industry launch. "If Colorado farmers get some seeds in the ground and are successful, that'll also really help the legislative situation in the heartland," he said. "This is going to be fun to watch."

But Hermann, even when worked up by prohibition's insanity into what struck me as her quite intentional and effective hillbilly mode, is at core a sensible, sanguine midwesterner. When I asked for her advice to American farmers, the first thing she said was, "After legalization, take it slow. Don't expect to reap a bonanza from your first crop."

From there she went straight into heartland schoolmarm mode. "Make sure your harvesters are sharp—know how to operate the combine so it works well with hemp . . . Dry your seed immediately to 8 or 9 percent [moisture] after harvest to protect the essential fatty acid profile during storage."

Useful intelligence, I'm sure, but not exactly the cheerleading you'd expect from someone the Canadian government considered a "Unique Skilled Worker" (essentially bestowing permanent residency on her) before finally making her a dual citizen. And there's a reason for the business-like posture.

"I want to see hemp just be another ingredient in a farm economy . . . providing healthy food structure and industrial material for people," she said. No politics, in other words.

HEMP PIONEERS

Mark Reinders, HempFlax Deputy Director, Oude Pekela, The Netherlands

The meadows of northern Holland were still frosted when I set off on an autumn morning to visit the nearby HempFlax headquarters in Groningen province. Perhaps the coolest part of my research for this book—and that's like choosing between favorite ice cream flavors—came very near the end, on the HempFlax factory floor. That's because I found myself watching (and in turn touching) the actual hemp fibers that go into Mercedes and BMW door panels. These emerged in clumps from a mechanized separator that sent the remaining hurd down a different chute (for use as cat litter).

Operating like a page out of the 1938 *Popular Mechanics* article that hailed hemp's twenty-five thousand uses, HempFlax also sells parts of the European industrial cannabis plant harvest for textiles, paper, and building insulation. The vast, noisy factory I was touring this chilly morning churns out more than 1,400 pounds of hemp fiber every hour.

Even though the company does four million euros business every year, its boss, 32-year-old Mark Reinders, told me that finding markets for the locally harvested hemp is "like juggling—we sell the bast fiber and then have to find markets for twice as much of the shiv."

On the agricultural side, the business requires constant innovation, too. Reinders pointed to a giant harvester parked next to the factory and said that his mechanics still have to jerry-rig equipment to fit a particular field's dual-cropping needs.

"See here?" he said, hopping up about eight feet to the harvester's hood. "We welded a forklift mast up top here so we can harvest the leaves and flowers higher up on the plant," he said. "That gives us a market for juice and shakes before the main blade cuts the stalks down at sixty centimeters to begin the fiber-retting process." Hey, presto, another kind of dual cropping invented. I was blown away that there's still no standard hemp-harvesting *modus operandi*, even in the relatively mature markets of Europe.

I loved that all of the hemp for the HempFlax factory has to date come from surrounding farms on the Dutch countryside, but Reinders said that price competition from GMO corn has forced the company to buy farmland in Romania as hemp fiber demand increases worldwide. This really gets his goat. "It's being grown for inefficient energy, not food," he said. "It's ridiculous."

He hopes the high corn price problem is temporary, because Europe's soil needs hemp. "I came to hemp because my father's a farmer and he cultivated it in 1996 as a cash-providing bridge crop that was a soil restorative," Reinders told me back on the factory floor as I snatched an armful of the most combed, highest-end bast fiber from the end of a factory conveyor belt (it was so soft to the touch that I felt like I was squeezing silky air). "I liked how fast it grew and that it was pesticide free. So I interned here as an industrial engineering student, and after I graduated from business school, (company founder) Ben (Dronkers) brought me on."

The privately owned HempFlax, which Reinders described as "on its feet, stable" and on a twenty-year, uneven climb to consistent profitability, was already supplying European automakers by the time he came on in 2007. "The way that happened is the template was already in place for natural fibers like flax in automotive components," Reinders said. "And a combination of its fiber qualities and market forces made hemp's position progressively stronger. We should thank the auto parts contractors as much as the auto companies. It was the parts suppliers who were looking for affordable quality to keep their own costs low."

When I mentioned that North American hemp farmers have no modern experience taking care of a fiber harvest, Reinders nodded gravely and agreed with Canadian hemp researcher Simon Potter that we were talking about a vital body of knowledge that requires expertise. "We actually go to the fields to do the harvesting rather than letting farmers bring in the harvest," he said. "With fiber, the motto is 'quality in, quality out.' A farmer might be worried about rain and want to end the retting when the fiber is still gray. He has his own priorities. We come in and say, 'Wait three more days.' You want the fiber to be a dark yellow for the high-quality applications like textiles and industrial components." And so the North American hemp fiber learning curve begins.

CHAPTER ELEVEN

Teach Your Regulators Well

*F*or hemp to once again take off in the United States, history tells us that two more elements have to fall into place. First, the industry pioneers must work with regulators to craft domestic standards. I learned this from the saga of American biodiesel pioneers Kelly and Bob King. They were in biofuels so early, their Pacific Biodiesel website is biodiesel.com.

According to *Business World Magazine*, Pacific Biodiesel shared its pre-launch study results with regulators and even competitors because the world frankly didn't know how to make an industry of waste restaurant oil. Today their oil fuels a good deal of Hawaii, and they consult the world over. You can fill up at gas pumps on two Aloha State islands, and municipalities use the fuel for backup generators.

Similarly, the initial Canadian hemp players, several of whom are still in the industry, worked with regulators on everything from field-testing hemp varieties to THC analysis, right from the beginning. As we've discussed, this actually started several years before Canada's official 1998 reboot.

As Hermann put it, "Even if President Obama and Congress legalize hemp tomorrow, there's still a lot of work ahead for the U.S. market and anyone who wants to be a player."

The initial U.S. state hemp legislation generally nods toward the Canadian model: Colorado, in addition to unlimited commercial cultivation for registered farmers who grow hemp with that inert 0.3 percent THC limit, is making a vocal statement of top-level support by allowing those ten-acre developmental test plots wherein THC levels won't be tested until a cultivar is ready for the commercial market. Similarly, Hawaii's step one looks to be a hundred-acre state-sponsored research project. Pacific Biodiesel's Kelly and Bob King are big supporters of that project, because, in the end, the french fries that today drive their business are finite.

"Hawaii is close to legislation allowing for a test hemp plot that we hope will remediate a few centuries of sugarcane monoculture soil and provide energy feedstock," Kelly King told me.

Now, patiently developing a regulatory framework and official cultivars would seem to be essential. But there is another fairly loud opinion out there, and I'd be remiss not to mention it. It goes like this: The original American hemp farmers planted what they had on hand in their wagons after crossing wild rivers and unnamed mountain passes. And they managed, before interstates, let alone NAFTA, to build a world-leading industry.

In other words, some hemp activists make the case for starting now with that ditch weed (or, if you prefer, the "heirloom cultivars") easily found out by the railroad tracks in the heartland. This *Let Darwin choose what we plant* philosophy is running up against the *We live in a lab-coat-and-hairnet era because of uniformity and product safety demands* line of thinking.

Hermann's view on this comes with too much in-the-field experience to ignore, and it's basically this: Once she's expanded beyond selling carrots at the farmer's market, any farmer has to be savvy about her choice of variety.

"Every Walmart already carries hemp oil, Nature's Path hemp cereal, and hemp twine," she said. "A mature industry has to be ready for the professionalism that level of reach demands."

She's talking about standards, testing processes, and certification paperwork. Humanity's oldest plant is about to grow up. "We have food and health inspectors certifying our industry in Canada," she reminded me. Burritos in front of the Phish show this is not. Still, this first to-do item is standard business stuff. It can be easily checked off.

The second crucial hemp industry to-do item sounds a little woo-woo at first, but it comes down to the very bottom-line concept of "preserving the brand." I firmly believe that if industrial cannabis is to emerge in a big way Stateside, its initial large purveyors must stick to the root values that, over the difficult decades of prohibition, have guided the industry: an industry that without question has also been an activist movement.

What are these core values? You only need to look at how the hemp industry standard-bearer, Dr. Bronner's Magic Soap, operates as a business.

First off, this Ivory Soap of the American organic food co-op is a third-generation family company—older than that, really, since the Heilbronner family was making soap in the old country before emigrating to the United States and dropping the *Heil*. But today's CEO David Bronner is the grandson of the famously gifted if verbose modern founder (as many of us first discovered in a

HEMP PIONEERS

David Bronner,
CEO of Dr. Bronner's Magic Soap

Who knew? You can be president of a fast-growing, multimillion-dollar, family-owned company while sporting a ponytail, mentioning Bob Marley's birthday on your company's official website, and stating openly that you like "all forms of the cannabis plant."

The successful can generally live how they like. Yet in an era that's demoted "prominent businessman" to a social rank just below "used car salesman" and just above "congressman," thirty-eight-year-old Bronner wears the mantle of a solidly old-school hero. That is, to the type of patriot who believes prosperity is best engendered by running a sustainable business that shares the wealth with its employees. Though he deservedly gained attention by risking his own freedom to ensure that his company's sustainability ethic remains undiluted (this when he locked himself in a cage in front of the White House in 2012 with only a potted hemp plant for company), Bronner's greatest accomplishment to date might have come while working within the law in 2004.

That was when he coordinated and helped fund a coalition of hempsters who pulled off a Lake Placid–level miracle: Helmed by Hemp Industries Association lead attorney Joe Sandler, they defeated the DEA in federal court (that was what happened on Bob Marley's birthday, by the way). This after a protracted legal effort to prevent the federal banning of hemp food products from import into the United States. It's one of the great unheralded victories of the drug war.

"We were already using hemp," Bronner told me. "When the Bush administration began seizing hemp shipments at the Canadian border, we sued to be able to keep using it."

The 2013 U.S. House passage (and very likely 2014 full congressional passage) of the first federal hemp cultivation law in half a century is a direct result of that underdog victory nine years earlier: One of the most compelling arguments hemp supporters like Representative Jared Polis (D-Colorado), Senator Rand Paul (R-Kentucky), and Senator Ron Wyden (D-Oregon) were able to make to colleagues is the absurdity of importing a crop that the U.S. farmer could be profitably growing if not for Big Government regulation.

In Wyden's words, "The outlandish restriction on free enterprise hurts rural job development and increases our trade deficit." If Bronner and the rest of the parties to the DEA lawsuit hadn't drawn a line in the sand in 2004, we might not be on the cusp of a new, billion-dollar domestic agricultural and manufacturing industry today. To be able to import and use but not grow a crop is, as farmer Bowman put it at his Fourth of July hemp planting, exactly the kind of absurd trade restriction that fomented the original American Revolution.

So how did Harvard alum Bronner come to hemp activism? "In Amsterdam some years back I had some frankly intense revelatory experiences while using psychoactive cannabis. Those experiences convinced me, after years of dismissing it, that my grandfather [company founder Emanuel Bronner] was right when he spoke of a spiritual unity and a quest for world peace."

friend's bathroom late at a party, the company's bottles are covered with spiritual, prophetic, poetic, and philosophical citations). The important part is that even with fifty-three million dollars in sales in 2012, the Escondido, California–based company is still run by a spiritually minded and progressive family, and this is reflected in its policies.

David Bronner, like all the company's executives, makes only five times more than the company's lowest-paid employee. He also makes sure the company buys some of its olive oil from an orchard worked jointly by Israelis and Palestinians, and uses only organic, fairly traded ingredients.[43]

Furthermore, the thirty-eight-year-old Bronner, when I emailed him a few follow-up questions to a recent phone call, was in Washington State fighting GMOs. He's also led a legal charge against greenwashing body care companies that dangerously call their products "natural." Not only that, Bronner is an American agricultural patriot. Eager to reduce the twenty tons of Canadian hemp seed oil the company imports annually, Bronner told me in February 2013 that the company is "financing a project to collect and develop cultivars for American latitudes and soil." Six months later, as we'll see in a little while, I visited that project.

If all that weren't enough, the reason Bronner added hemp to the family recipe in 1999 was pure performance: "It makes a better emollient," he told me. "Less skin-drying." At the same time the hemp went in, caramel coloring, in Bronner's view the sole unnecessary ingredient in the soap, went out. This was, in the end, an artisan soap maker improving the generations-old family product. As a before-and-after patron who washes my kids in the stuff, I can attest that the hemp version is demonstrably superior. You no longer need to dilute it before using it on children's skin. The point is, Bronner didn't add hemp as a political gesture. He added it as a customer service move.

This is a company model America can get behind. Not just because it's righteous and so much less evil than, say, some satellite dish companies or banks I could name, but because it reflects values that are good for America's economic and spiritual future.

Consumers today associate hemp with concepts like admirable ideals, healthy cooking, and the world's strongest rope. Ask Arm & Hammer if it's important that we associate baking soda with non-toxic freshness at an affordable price. Good reputation is bankable in a cynical age. So you might say Bronner and Co. are enjoying karmic payback for fighting to institutionalize industrial hemp in the United States.

The bottom-line benefits of this "Don't Just Not Be Evil, Be Actively Good" brand are not just my opinion. The 2002 paper out of Purdue University that called hemp a "camp follower" plant also concludes that hemp is "pre-adapted to organic agriculture, and accordingly to the growing market for products associated with environmentally-friendly, sustainable production. Hemp products are an advertiser's dream."[44]

Speaking as a consumer, I can say without hesitation that if farmers organize vertically in such a way that I have a farmer-direct option for my hemp oil, ideally a local one, I will gratefully take it.

Will Big Ag move into hemp? Almost certainly. That's why Canada has its hemp GMO ban in place before there is GMO hemp. For its own good, U.S. industry players would do well to do the same. Franken-hemp? The public would go elsewhere. I mention this because North Dakota's argument for immediate legalization has included the bullet point that University of North Dakota researchers are developing a sort of glow-in-the-dark gene that can serve as proof that a hemp plant isn't a psychoactive variety. Very Bad Idea. Bye-bye featured shelf at Whole Foods. GMO tainting erases hemp's strongest selling point. Plus even the inevitable McHemp sandwich needn't be genetically modified. The plant grows like a weed.

The fact is, venture capital is already flowing into cannabis because corporate bean counters see a market. The "sums" add up. That's what winning the drug war looks like, I'm afraid. Wall Street does what Wall Street does. But you'll soon be able to support your local hemp farmer and buy her oil at the farmer's market and food co-op, as well as at Walmart.

Support Your Local Heartland Hemp Homesteaders

*A*merica's small farmers are becoming increasingly aware of the new old cash crop awaiting them. A year ago, third-generation Colorado wheat farmer Jillane Hixson never imagined she'd be talking about hemp. She in fact wasn't sure about the difference between hemp and psychoactive cannabis.

Now, thanks to Colorado's cannabis legalization and despite her own conservative county's proclamation that it doesn't really like the new cannabis constitutional amendment that 55 percent of the state's voters passed in 2012, Hixson told me with a point to her drought-crippled fields on the morning I visited, "Hemp certainly is intriguing to me. We have to try something different."

She was referring to the fact that hemp's wide climatic adaptation and fast-growing foot-long roots allow it to thrive in drought-damaged soil. Indeed, cannabis worldwide traditionally has been used for erosion control.

"My daddy planted it along the ditch, so it wouldn't flood," a Nebraskan farmer named Tammy told me. "Then in the fall we'd graze the cattle out there: They sure loved it."

HEMP PIONEERS

Jillane Hixson,
Eastern Colorado Wheat Farmer

I met Hixson in 2012 when I gave a talk about cannabis in Lamar, Colorado, a place where local Democrats call 30 percent election showings "moral victories for our message." Her family's been in the area for generations, but after the talk, she showed me her farm and said, "Something's different now."

Right on the driveway as soon as I stepped out of my truck, a handful of beach sand (formerly known as topsoil) blew across our ankles. "This isn't what the soil is supposed to look like," she said. "They've even tried dumping municipal waste for New York City on it. It just washed away."

Desperation was opening Hixson's mind about hemp. "We planted this wheat in the fall," she said, pointing to a stretch of cinematic desert that was supposed to be amber waves of grain. "It should be a foot tall by now. We just haven't had any precipitation at all. Again. The wind tears up the entire field."

This was original Dust Bowl country in the 1930s, and what Hixson was saying is that it's back for a sequel. Hixson and husband Dave were stranded in their home for half a day during a May 24,

2013, Depression-evoking three-story dust storm. So much dirt was flying into the house during the overnight disaster that, Hixson told the *Denver Post*, she and Dave had to cover their faces with handkerchiefs and "sleep with their heads under blankets."

I remembered this when, a year after I'd visited her farm, a federally written Dust Bowl exhibit at nearby Comanche National Grassland ascribed a good deal of the blame for the first Dust Bowl to "disastrous land management practices."

I dunno. Sounds a lot like modern monoculture. That's why Hixson's opened her mind and calls hemp "compelling." "We have to plant a crop that will take root and hold this highly erodible soil," she told me.

These are the eastern Colorado plains, a place where, in topography, demography, and percentage of linebackers produced per capita, you *are* in Kansas anymore. It's not Boulder, is what I'm getting at. Hixson is no flower child. The thing is, you could shoot a French Foreign Legion movie here, so Sahara-esque is the parched countryside.

In fact, just as I was letting a handful of Hixson's beach volleyball sand strain through my fingers like an egg timer, husband Dave materialized and chimed in somewhat emphatically, "Look at this wheat crop. It's a dust bowl out here . . . Sure we'd grow hemp if it paid. If there was an outlet to sell it."

That's what farmers always do: They adapt; they look for a new crop that will make them money. Since, as we've seen, so many of those outlets Dave was talking about will be industrial, we'll leave the farm and give the final word to the Composite Innovation Centre's Simon Potter. I do this not because his point of view concurs with both this book's thesis and my dreams, but because he is, in an Aristotelian sense, a reliable source.

That is to say, Potter's genuinely not a political hempster. He gets paid no matter what the best fibers turn out to be in the lab. And they're turning out, in areas ranging from roofing to insulation to aircraft bodies, to be industrial cannabis fibers.

"We've been cultivating this plant for eight thousand years," he told me next to the hemp tractor. "[Now] we're rediscovering its manifold benefits."

You had me at *eight thousand years*, pardner. Will the mainstream American marketplace really embrace all the hemp housing, food, power, and tractor parts we've been discussing? We're going to find out soon enough. I think yes.

If the Canadians are right about their seed oil sector's growth rate, and the Europeans are correct about hempcrete's market readiness, Jack Noel, who co-authored a 2012 industrial hemp task force report for the New Mexico Department of Agriculture, says we're talking about an industry that will be the fastest in U.S. history to reach the fifty-billion-dollar-per-year mark.

After what's turning into eight months of in-the-field investigation, I can report that I, too, believe that in a decade and a half, industrial cannabis will outearn even its lucrative psychoactive cousin in the United States. Here's why: Coors is big. But ExxonMobil is bigger. How much bigger? The *Denver Post* proffered on January 14, 2013,[45] that industrial cannabis could be an industry as much as ten times larger than that of psychoactive cannabis, itself already America's number one earning crop.

Perhaps more important, the simple, ancient cannabis plant provides, after industrial harvest, a residual feedstock for regional-based sustainable energy production that cuts out at once Monsanto, BP, and Middle East oil dictators. And it gets out Ring Around the Collar.

So I hope you'll join me in wishing America's hemp farmers well. Or better yet, become one, or a partner in a processing plant or a member of an energy cooperative. These are the people who are going to be reducing GMOs in the industrial food system and in so doing helping heal our farmland from monoculture. They're heroes, in my book. I hope they will be numerous and prosperous. This is the health of my offspring we're talking about.

Said replicants currently napping, I'm off to bind a batch of tax receipts with Romanian-made hemp twine. Funny: The cannabis plant (care of the publishing and journalism games) has earned me the bulk of my income this year. I wonder how many years it'll be before this twine, which I got at Walmart and which lasts years even out in the Land of Enchantment sun where it also secures my grapevines, will be available as an American-grown product? I guarantee you it'll make quarterly tax time a little sweeter that year. It might happen sooner than anyone thought. Like this year.

You know what? Forget taxes. With the bright thought that we humans still have both truly sustainable and lucrative options in our climate stabilization arsenal, I'm going to heed the summons of the hummingbird hovering outside my office window and head out for a run up my canyon, powered by the King of Seeds.

Watching Cannabis Displace Corn in the First Digital Age American Hemp Fields

Note: For publishing world reasons, I was bestowed a few more months to work on this book than I thought I would have. That thrilled me, because it allowed me to keep at the research until today, a little past midway through the 2013 U.S. hemp-farming season. The rest of the book results from that fortuitous turn of events. —DJF, August 25, 2013

S ince I was at the moment watching a country band fervently belt out music of the *Show me the birth certificate* variety in a conservative farm town in rural Colorado, while perhaps a hundred yards away a ceremonial hemp planting was about to take place, you'll understand why I say that July 4, 2013, was one of the few times I've ever had to actually pinch myself. (Another was when, as a congressional intern in 1988, I heard George H. W. Bush announce that his running mate was a fellow named Dan Quayle.)

Nope. Even though there were other factors contributing to the fantastic—and for a journalist covering a forty-year-long drug war, almost phantasmagorical—qualities of this day, this was not a hemp-ster-patriot's daydream. Everything was really happening. To name one thing, at that very moment sixteen hundred miles to the east, an American flag made from hemp was, thanks to the fellow alongside whom I was eating jambalaya and ignoring lyrics, flying atop the U.S. Capitol. This must've been how the average Vermonter felt when word filtered north that the colonies had actually won the Revolutionary War. "Really? So fast? Against such a better-funded adversary?"

And the Founding Fathers were on our fifth-generation farmer Michael Bowman's mind. He was the reason for both today's hemp planting (he chose the day and the location) and the Old Glory hoisting in DC (he lobbied his congressman—more on that in a little while). Here in the economically challenged farm town of Byers, its population of eleven hundred down 6 percent since 2000, Bowman, in front of my (and a Denver ABC television crew's) eyes, was going to make a hemp statement that he called "a shot heard 'round the Beltway." He wanted a meaningful day to commemorate the start of what he routinely calls "a new agricultural system."

This was taking place on land owned by his longtime and very conservative rancher friend Gary May. Bowman was going to plant one of the first American hemp plots since World War II on a tiny patch in the back forty of the May Farms big tourism operation, in fact. "Colorado's number one agri-tainment destination," May's voicemail tells us. That means things like tractor pulls, weddings, and Replacements concerts. The hemp was going to share the space with corn, not far from the May pumpkin patch that, along with hay bale rides, highlights the autumn-time May Farms Harvest Festival.

"In every major movement, there has to be people ready to lead by example," Bowman had told me an hour earlier as he pulled out a five-hundred-gram package (around four cups) of European hemp seed from my rental car in the crowded May Farms dirt parking lot—which was adjacent to a stables. "That's what we're doing today. Our own Independence Day two and a quarter centuries ago started with a shot. We're leaders in Colorado on this issue. We know we need to make a transition to a new farming mind-set, in order to make a real difference for our grandchildren—for my own grandchildren. That's our ammo: planting hemp in a GMO cornfield."

"Hemp displacing corn," I recall musing dreamily as we shuffled with the crowd inside May Farms' vast event facility. This was a big metal barn-like structure that can hold a thousand, easy. From the outside it looked like you could host a rodeo. Inside it was comfortable, with picnic tables feathered around the dance floor.

"Now you know my master plan," Bowman said. "The transition back to the crop of our forefathers."

"You're a conservative," I observed.

He laughed and looked at his bag of smuggled seed. "Yes, I guess I am. Oh, look! Jambalaya."

This was to be a symbolic planting Bowman had decided on for 2013. Headline: *Hemp Seed Planted on Independence Day*. Not in headline: *Actually a very small plot that might or might not be tended through the season.*

Part of the reason Bowman decided to go this route (he had until very recently been telling media, including me, that he aimed to plant a hundred acres this year) was that his octogenarian parents still planted corn, wheat, alfalfa, and other rotational crops at the family farm in Wray, the one we earlier discussed. These crops were connected to multiple USDA programs that, in light of those direct DEA threats against Bowman, he didn't want to put in jeopardy. At least not prematurely.

Which leads to the other reason he decided to wait—as even most bit-chomping Colorado hemp farmers were waiting—until the 2014 season to plant a full crop. The state's hemp cultivation advisory committee—with the support of the Colorado agriculture commissioner—would be instituting its legislatively mandated commercial planting regulations starting with the 2014 season. In fact, Bowman was on that committee.

"As far as the state is concerned, next year Coloradan farmers can grow legally and commercially, without the need for federal approval," he told me in the jambalaya line.

Whatever the reasons, Bowman could hardly have chosen better than this spot an hour east of Denver. Picturesque down to the classic grain silos, Byers, Colorado, like much of the America heartland, is literally dying for a crop to rescue it from its drought-inspired decline. Local patriarch May's property was the site of the town's Independence Day celebration. Everyone was here—five hundred people easily at any one time—and by the time of the planting some were still conscious.

I've never seen folks mean their Fourth revelry so deeply. As they made their way between the refreshments tables and children's train, they wore American flag shirts and giant foam stars-and-stripes cowboy hats the way college sports fans paint their faces in the team colors on the day of the Big Game. Myself, I was expressing my love of country by showing up entirely clad in hemp: collared shirt my Sweetheart made me for my birthday and Canadian-made hemp pants. My sun hat even had a hemp band.

This was a very specific America being celebrated at the Byers bash, comprised of a demographic that chants "U! S! A!" at its Olympic broadcasts. I felt like I was in a *King of the Hill* episode. Yet when Bowman told him what he was doing at the shindig, our jambalaya chef Mo, inventor of Cajun Queen and Papa Mo all-purpose seasoning, said, "Why are you planting so little?"

This attitude wasn't just typical of what I heard from the Right side of the political spectrum at the Byers July Fourth celebration; it was reflective of the nearly unanimous view. That a pro-cannabis viewpoint is proving not to be a "pinch me" reaction to hemp anywhere in the heartland in turn felt journalistically significant to me. It's contrary to cannabis's reputation. Heck, right as Mo was giving his pro-hemp shpiel, the band was playing "Okie from Muskogee."

The conservative embrace of the plant is, in the end, why the war on cannabis is over. It's one of the few things nearly everyone agrees on. Since when do 80 percent of Americans agree on anything, as they do that the drug war is a failure? If President Obama stepped to a podium tomorrow and said, "Hemp hemp hooray, the Digital Age Homesteading Act goes to Congress tomorrow," his approval ratings would go up. Nationwide. Kentucky might even go Democratic.

In other words, watch for big policy change when MoveOn and Tea Partiers and everyone in between agrees on an issue. For his part, the sixty-five-year-old May was what everyone told me was characteristically garrulous about his ol' liberal buddy's ceremonial planting that day. He spoke as an agronomist.

"I am an eastern Colorado farmer," he said, joining Bowman and me for jambalaya after parking the golf cart he used to oversee the day's festivities. "I need to take two looks at my crops: the economic and the agronomic view. The agronomic view is the approach of another rotation crop that might help me become more productive in an area of the world that gives me no promises of water. When I look at my thousands of acres of dryland wheat, my thousand acres of corn, I need to consider all opportunities. This might be an opportunity to make something work economically, especially if we can get some processing going nearby."

In the midst of a very Toby Keith–heavy set, I asked if May cared at all about drug war rhetoric. "I try to stick to the agronomic," he said. "I'm trying to produce food for the people of this country. If hemp moves into the rotation for me, I'm more than willing to plant it. So we'll see what happens in Colorado in this next year. If there's really a market."[46]

I sat back in my chair to listen to what was probably the nine thousandth lifetime policy discussion between Bowman and May, both people who can get along with anyone. It wasn't long, though, before the political talk turned to the goings-on in the actual dirt— because that's where the facts reside. Rainfall levels don't lie. In fact, one of the coolest parts of the day for me was listening, as a suburban-raised recent convert to Southwest ranching, to these two fellows, who had local farming in their blood for a combined total of ten generations, talk shop.

"This cultivar'll take a twenty-seven-degree freeze, which is nice," Bowman observed of his seed.

"I like that it doesn't have to go in in March, like canola does," came back May. "I mean, June first, you're almost in there with millet."

"He's trying to keep a struggling community alive," Bowman told me after May sped off to put out a fire surrounding the classic car contest judging. "This was Colorado's first county. If Gary decides to plant a full hemp crop himself in future years, and I think he will, a lot of people will see it. He's got a corn maze, and pumpkin picking in the fall, and as you can see we're right on the freeway—a tall crop would probably be visible to drivers."

So *that's* why Bowman chose this planting site. "Gary thought, *Hmm, I wish there was a crop that could provide income while helping us deal with drought*," I said, chewing contemplatively.

"And he came to hemp," Bowman said. "At least he's considering it. Which is why it's so astonishing that what we're doing here today is so political. It shouldn't be."

I picked up the bag of seed. "Doesn't look like a felony," I admitted.

Bloated with southern cuisine, Bowman and I shuffled out to May's back forty after lunch and a brief tour of the crises-ridden classic car show (calm appeared restored). We came to the site of the hemp ceremony—near the middle of a rectangular field at the moment home to ten acres of somewhat sad-looking corn. I scooped up a handful of grainy dirt. "I keep thinking *Sahara* when I see the heartland topsoil these days," I said.

"Welcome to the Central Great Plains of the United States of America in a time of climate change," Bowman said. "We're in the middle of historic drought. All the crops are drought-stressed around here. Our aquifers are drying out, too, so we can't pump the water the current crops need."

"*Drought-stressed* sounds a little euphemistic," I murmured, dropping the hot handful of GMO-worked dirt. "This corn looks sick. Like *stay home from school* sick."

Bowman shrugged. "This is the new normal. It's a serious challenge before us, and we're here today to find solutions.

"This," he added, peeling open the seed bag, "is a rescue operation. A recovery crop. Replacing, as you say, corn with hemp."

It's a recovery operation in the field and in the history books, I thought. Because we're talking about reclaiming the mantle of the early independent agricultural principles on which the United

States was founded—in the name of sustainability. Bowman was saying that a locavore, GMO-free, and yet lucrative economy is the most patriotic model for the future.

He made his way to the heart of the field, looking comfortable among corn rows, which I observed aloud.

"I've grown a lot of corn in my life," Bowman confirmed. "Including a lot of GMO corn. I was cornfed." As most of us are these days.

"No planting machinery?" I asked.

"We're doing it the way our forefathers did before automatic seeding," he said, digging his paw in the seed bag for a handful of what could, in a decade or two, be among America's most profitable crops. "We're scattering it, and then tamping it down for good seed–soil contact. I'm well-versed in the technique from my family's own ranch."

And so the time had come. "This is a crop to be rediscovered and used to create a new twenty-first-century economy for America" were Bowman's preparatory words.

"I'm going to let that classic car exhaust fill the role of the John Philip Sousa march that ought to be playing right now," I commented, then added, of his seed-scattering pose, "That'll be the statue, right there, hold it!"

I took a photo as the first handful of seed arced over some desperately thirsty corn. The motion conveyed a horseshoe pitch.

"First shot fired," Bowman said. It seemed right. We stood in respectful silence for a few seconds, a breeze making one nearby cornstalk appear to salute.

"How's it feel to be an official hemp farmer?" I asked.

"It's the most patriotic thing I've ever done."

"God bless America and everything," I said. "But let's face the facts. Technically that's a federal crime you just committed and I, as Peter Tosh puts it, am going to advertise it."

Bowman's beaming, guileless farm boy mouth sort of melted from its default grin for a moment. Then he sprinkled another handful of seed. "I guess it is a federal crime," he said. "The DEA has jurisdiction right now over a crop that should be under the Department of Agriculture, and they're worried about job protection in their agency. But that's one of our easier challenges. This justly rebellious spirit is how our country gained independence. We're going to create a new industry to transform agriculture. America's going to be stronger with hemp in her fields."

He looked down at the bag. "A few left—let's go back inside and throw it into a protein shake."

Still, I wondered aloud, "Why do this so publicly? Why flaunt federal law with a couple hundred seeds that probably won't be harvested, very shortly before, but still before, the federal government is ready?"

"This is about freedom, about gaining independence from federal farm programs that might have had their purpose, but which keep us ranchers chained to a very specific set of crops. We're talking about opening markets, and all kinds of legitimate commerce that'll result. It's long past time that American farmers be allowed to do this."

Bowman is polished enough in his soundbites, but I saw where his greatest strength lay this day after we headed back inside the event building: making the case to folks who listen to a lot of Lee Greenwood. Over the course of the next two songs, he worked the room, dance-walking. I watched as, one by one, he pressed palms, very much the hemp politician, winning over good ol' Americans to hemp.

"Five hundred grams of viable hemp seed, that's what we planted today," I overheard him telling a young woman pushing a stroller. "It's from France, that's all I can tell you. It came from a farmer friend over there. It's a little cooler there, but this cultivar should be fine here."

He'd made the same point to the Denver television crew, trying to open farmer minds, media minds, all minds. Since Bowman kept making the economic case for hemp, it occurred to me that, on top of all the practical reasons, Byers is also a homonymically apt name for the location of a patriotic planting. That's because as soon as you harvest this crop, there are indeed buyers. Have I hammered home the point that there is not nearly enough supply to meet hemp seed demand in North America? Right now.

"You like hemp preachin', don't you?" I asked during a quiet moment at May Farms (I think the song playing was "God Bless the USA").

"It's effective because it's all true," he said. "I've found that anytime someone gives me five minutes, and I get to discuss the facts, hemp's role in the founding of our country and where we're going next as a nation, that person is a convert. I think I'm batting a thousand on that. When I talk about Henry Ford growing a car from this American fiber that's stronger than steel, and fueling it with ethanol from the same crop, this speaks to people."

To say that Bowman's a busy guy these days is a severe understatement. Ever since he had that grand success as an activist ushering in Colorado's landmark renewable energy mandates in 2004, he's been a full-time sustainability lobbyist at the state and national levels—U.S. senator Mark Udall (D-Colorado) calls him The Human Hand Grenade. And it was while working on hemp's U.S. House passage late last spring that he had the inspiration for the hemp flag. After the Denver TV crew left Byers and as we prepared to do the same, Bowman reminded me that the hemp Stars and Stripes was at the moment flying over the U.S. Capitol in Washington, DC.

"Seemed to me one of those irreversible gestures," he said of putting the bug in Congressman Jared Polis's ear to make the flag hoisting happen—congressmen are always shooting up flags for constituents to honor someone or some event. "I was in the House gallery when the hemp amendment passed. I thought that once a hemp flag flew, it couldn't be un-flown."

In the last moments before we'd pulled out of May Farms just ahead of the nearby fireworks that remarkable Independence Day, I recall watching Bowman, as is his wont, maybe even his unique specialty, shift comfortably from Beltway lobbyist to local rancher.

"The DEA doesn't need to lose jobs on hemp," he told Gary May and me over one last bowl of Mo's spicy work of art. "They can help us enforce the industry's regulations from a supportive standpoint—crop inspections and testing, farmer registration. Just like Canada did with its Mounties. I consider my job to be convincing Congress to change the DEA's prism to one of being part of the hemp industry. I'm an optimist. Even their position will evolve."

Then, not ten minutes later, I heard Bowman call out his parting words to May's teenage son, Grant, as we hopped back in the car. "Those hemp seeds'll sprout in about three days," he said, leaning out the window. "Keep an eye on 'em if you can."

A Dust Bowl Antidote— It's About a Cash Crop in Today's Soil

*F*ast-forward a month. Across the state of Colorado from Byers, 256 miles to the southeast, an agriculturally signif- icant 2013 hemp planting was well under way. Nothing symbolic about this one. We're talking sixty acres, intended to help build a domestic seed stock for both oil-producing and fiber-pro- ducing varieties of industrial cannabis.

As I, traveling with my family after some days spent unplugged in and around (mostly in) wilderness rivers, bumped across Comanche National Grassland on August 11, 2013, I reflected on how hard it is to get lost in Colorado, even for me, who can get lost in my kitchen. That's because the mountains are to your west. Pike's Peak. The Rockies. The spine of North America.

Heading east, as we learned during our visit to Jillane Hixson's nearby ranch, is a journey into a world so geographically and cultur- ally different from Denver's mile-high one that by the time I crossed into Bent County I felt I knew what interstellar travel will be like.

And that's explicitly why forty-year-old Ryan Loflin was planting hemp so far east. So far east, in fact, that during one of those plains country semi-naps one takes after setting the cruise control, I think I briefly crossed into Kansas. He wanted to show his childhood neighbors that hemp was an answer to the long drought that has savaged farm economies from Nebraska to New Mexico. A Dust Bowl antidote.[47]

We were meeting in Springfield (population 1,454, incorporated 1887), seat of Baca County, which in addition to Kansas also brushes Oklahoma and my own state of New Mexico, with Texas in shooting if not shouting distance. They should call this region the Five Corners. Denver is five very long hours away. I got stopped by two flash-flood-related road closures just trying to navigate between these distinct biomes.

"Everyone here is desperate for a viable cash crop," Loflin told me when I met him downtown. "They're hurting from this half decade of drought and looking to diversify."

Springfield is a town that keeps itself spruced up—the small grassy downtown park has a tasteful trickling fountain and rocking metal benches, and there's a well-maintained town Olympic swimming pool. Looking out the passenger window at the small, neat homes as we circled the original eighty-acre town site, my Sweetheart said it looks like a place that peaked in 1955. Or earlier: Named after their own Springfield by the Missourians who settled it in 1885, some of the buildings look like something out of *Blazing Saddles*.

To listen to him talk, you'd know Loflin was from here. He spoke with the almost southern twang of the eastern Colorado plains. To look at him, tall and thin and wearing a T-shirt featuring an in-over-his-head skier above the words JESUS SHREDS, you'd also guess, correctly, that he'd gone west. That he'd flown the coop to the alpine side of Colorado's intercultural divide for a time. That was not a Baca County–made T-shirt.

Because Loflin knows both worlds, he was in a position to inform me, in our first minutes of conversation, that I'd found myself in yet another place where the prevailing opinion about where President Obama was born has no bearing on the opinion on hemp. In other words, the need for a cash crop is erasing traditional cultural war boundaries. Or, according to Loflin, has already erased.

As I and my family loaded up to follow him the couple of windy miles to the twelve-hundred-acre ranch his family has operated since the 1930s, Loflin told me, "Oh, we've got some real conservative folks around here, and everyone's asking me when they can get some dang seed. So that's what I'm doing this year—trying to build a seed bank."

In pursuit of that locavore goal, prodigal son Loflin, who had until this year been running a reclaimed lumber business[48] and raising a family in the ski town of Crested Butte, was putting his sixty-six-year-old father's lucrative alfalfa operation, federally funded like the Bowman family farm, at risk.

"I'm invested in this community" was his explanation as I and my tribe piled into his farm truck for the short but bumpy ride from the Loflin farmhouse to its sixty-acre felony. "I have three cousins who farm here. I have my own kids' future to think about—my oldest is starting kindergarten. I knew I was coming back, and I want to create an economy. And I'm leasing the land to protect my dad."

Before we start throwing out the Gandhi comparisons, know that this is all part of an ambitious win–win scenario Loflin freely outlined for me as the truck rumbled up to the eastern side of the hemp field. His master plan, as the Rocky Mountain Hemp company, is to become nothing short of "the hemp seed oil expeller for Oklahoma, Texas, New Mexico, Kansas, and Colorado. I'm within eighty miles of all these states. Big picture, I want to be processing the region's [hemp seed] oil."

Hearing these words, I marveled that it was only five months earlier that I'd first heard people like Shaun Crew and Norm Roulet describe such a locally controlled vertical model. While the priority mission was to build seed stock this first season (which, remember, hemp authority David West told us at the very beginning of this story is so vital: without fail the absolute Objective One in the business plan), Loflin said he hoped that, on a very small scale, oil processing would start in 2013 as well.

"This is no pipe dream," he said, parking the truck in a sea of man-high, already densely flowering plants midway through their five-month growing cycle. It was about an hour before sunset. I hopped out, offered my Sweetheart a hand, and extracted my kids. The crop's hand-like leaves were doing Egyptian dance moves in the very slight breeze. The whole field was very quiet, except for the prairie dog squeaks.

"If I haven't found an oil press by harvesttime, David Bronner [CEO of Dr. Bronner's Magic Soap], who's contracting for ten of our acres this year, said he'd let me use one of the company's presses. It's important to get a seed-expelling process going now. That way as soon as the federal law is all dialed in and everyone has seeds, it's on."

Make no mistake, this was a professional farmer from a longtime professional farming family talking big-time professional farming. This was not a hobby garden plot. It was an agricultural operation.

"Our planting rig alone costs six hundred thousand dollars, between tractor and the air seeder," Loflin told me. The latter device features a turbo fan that precisely delivers hundreds of thousands of seeds at whatever density you set, replacing Bowman's horseshoe toss.

Big investment, big payoff, is the model in these parts, for any crop. If Loflin showed his father, let alone the rest of Baca County and Colorado, that the hemp crop is viable and has buyers of a value-added finished product, well, the senior Loflin might want to open up the rest of the twelve-hundred-acre homestead. Then we're up into the acreage Hemp Oil Canada's Shaun Crew believes is viable.

I did the math. At current prices, seems more like a jackpot than merely viable. Twelve hundred acres at $250-per-acre profit? Um. Three hundred grand. "Clear," as the farmers like to say. "Wheat clears thirty an acre," Loflin told me. You might add a zero to that three hundred thousand if the town processes and sells its own seed oil and protein cake. As with any new industry, it's all about getting the initial fixed costs handled, and the eventual per-unit price down.

Within the Loflin family, Ryan knew he was engaged in a bit of a *Show me the money* situation with the older generation—similar to Gary May's willingness to plant test crops and assess. Farming, to these folks, is pure business. Has been for generations.

Though Loflin said his dad was already totally supportive. "I've been working on his mind about hemp for ten years—he sees what Canadian farmers are making."

Listening once again to farmers speak casually of tracts of land that to this goat herder seem vast, I tried to imagine the economy of a U.S. county filled with twelve-hundred-acre industrial cannabis fields. It was overwhelming to contemplate. Want to know what my first impression was, as I, my Sweetheart, and my kids waded (well above head-high for the little ones) into America's first digital age commercial hemp field? Sixty acres is a lot.

With the three hundred gallons of biofuel that a 1975 University of Illinois–Urbana study concluded an acre of hemp can provide,[49] this one small test crop could keep eighteen thousand gallons of energy domestically produced every year. Devoted to protein, it will give forty-eight thousand pounds of seed.

It took about ten minutes to traverse the field of two principal cultivars Loflin was growing on the family land in 2013. One was a seed oil variety and the other a taller fiber cultivar. Our discussion about where the seeds came from was a tricky one, since too many specifics might get a provider in trouble. Importing viable hemp seed into the United States without DEA approval was currently illegal, though hopefully not once you're reading these words.

"The seeds came in on the down low" was how Loflin put it. The fifteen hundred pounds that made it through customs came via UPS. Two shipments were seized at borders. So we left it at this: The seed oil (often referred to as grain) cultivar came from Canada (it looked a lot like the Finola variety you see all over Manitoba), and the fiber cultivar from Europe. Back at the farmhouse near the family's watermelon and beans, Loflin was also experimenting with much smaller numbers of seeds he'd received from China and other parts of Europe.

"With both varieties this year, if we get a lot of seed production I'll be happy," he said. "That's unusual—in the fiber varieties it won't always be the case. I also wouldn't mind if there's some cross-breeding, and we end up with a variety that can provide seed oil and fiber."

He was talking about dual cropping, which you'll remember all the hemp experts consider essential. I couldn't believe how real everything was getting. I thought I was writing an optimistic book about the future; turns out I'm writing a practical one for today.

I thought of Anndrea Hermann's wish, that we come to find hemp simply a quiet if lucrative part of a healthy farm economy. I realized that her wish was already coming true. Seen at field level, hemp *is* just another viable crop for America's farmers. That's why when the federal ranger at my campground the next morning asked me what kind of work I'd been doing in Springfield, I said, "Writing about the farm economy."

Loflin and I stood during what photographers call the magic hour, the sun just barely a full molten circle, at the spot where the two hemp varieties met in a line leading to the horizon. The flowers forming along the top of the fiber crop were a lighter, almost kelly green compared with its emerald leaves and stalks. Some of the blossoms were two feet long.

A fairly long silence ensued (not counting the cries of "I'm a prairie dog!" from my five-year-old, invisible but very audible around my ankles). Given that he is pretty much the only farmer U.S. growers can look to for actual in-the-field hemp-cultivation advice,[50] I asked Loflin how the debut season was going.

He laughed. "Well, hemp's about as hard or as easy to plant as any other crop," he said. "I'm really learning each part of the process as I go."

I asked for a for instance. Loflin laughed again, scanned the ground for about half a second, and snapped off a bushy piece of grass, about three feet tall. "Hemp may grow like a weed, but when you water a field, plenty of weeds grow like weeds," he said. "This is foxtail grass."

There was rather a lot of it. In places it was hard to tell where the blossoming hemp rows were, so gracefully were they sharing the space with other flora.

"Next year we'll plant thirty-inch rows instead of eight-inch, so we can run a cultivator between rows, do some manual weeding. Herbicides are out of the question—I don't want to and we can't for the Dr. Bronner's acreage—it has to be organic from the start. And I think the crop is looking good, for our main purpose."

Which was that seed stockpiling—one acre of hemp, remember, today yields about 800 pounds of seed (the world average is actually 875 pounds, according to the USDA). Each successful harvest thus means an exponential growth in the available seed stock. This kind of field was the genetics lab before Monsanto.

I agreed with Loflin that the crop looked and smelled very green and healthy. The terpenes (or resinous hydrocarbons) in hemp provide the distinctive, almost minty smell, and it bespoke fecundity. "The flowers are so seed-heavy," I noticed. "Some of them are losing battles with gravity. They're everywhere."

"Look at this one," Loflin said. It was a plant that, situated at the edge of a vast row, had been pelted into horizontality by recent hail—I myself had gotten pelted in the same front. "Look how it's still green and growing along the ground," he said of the bamboo-like stalk. "It's amazingly hardy."

"Ditch weed teaches us that," I agreed. "Seventy-seven years of eradication and it's still here."

For his next demo, Loflin reached a few rows deep into the fiber cultivar crop and gave a five-foot-tall male a vigorous shake. "Check this out," he said as a pollen cloud emerged that obscured the light for several seconds.

After I spent a few more minutes marveling and playing hide-and-seek with my prairie dog kids, Loflin directed my attention to a seed he was trying to pinch from the ice cream cone of a female flower on the oil side of the two crops we were straddling.

"This one's ripe," he said, hunched like a jeweler examining a gemstone. From a cluster of what appeared to be at least a hundred seeds in that one flower, he extracted the coconut-colored individual, a quarter of the size of an orange pip, and squeezed its already opening calyx. A lacy white seed cake emerged like a prize, floating in an oily emerald pool.

"That's where the omegas are," I observed, drooling,

"That's where the money is right now," Loflin observed, drooling. "The billion-dollar protein that's driving the market."

As the sun set, I shot video of row upon row of industrial cannabis plants that seemed to be maturing almost visibly. Mosquitoes were devouring humans and prairie dogs alike, but I for one didn't mind or indeed even notice my own bites until the coyotes were going off that night outside my family's campsite, a hundred miles away. I was that Revolutionary War–era Vermonter again: I was reading the news sheet in shocked bliss. I felt like I was seeing a bright economic future for my country, family, and planet for the first time in a long time.

On the planetary side of things, Loflin told me he was trying to prove via year-by-year nutrient comparisons whether industrial cannabis really helps heal drought-damaged soil. He was already demonstrating that it will grow in it.

"I'm sending off one of this year's soil samples next week," he said. "I can't wait to see how these compare to future years."

And so we were back, as all conversations in the western plains eventually return, to the encroaching Sahara issue. I broached the topic delicately, since Bowman had reminded me that folks in these parts consider themselves the caretakers of the land. But no one is denying that the Dust Bowl is getting real again here, in the same place it struck seventy years ago, right about when hemp got stifled. That Dust Bowl is in the collective memory here. It's also in the official memory. In fact I saw a photo from a 1935 dust storm in my previous night's campground, and it looked exactly like the one Jillane Hixson sent me that trapped her and husband Dave in their home for fifteen hours.

"How much do you think climate change plays into this ag crisis?" I asked Loflin.

"The land is not producing as it was ten years ago and prior to that," he said. "Because of the lack of rain. Nutrients are not being delivered. They're gone. But this is cyclical. This topsoil can come back, and I have no doubt hemp will help the process."

I and my Sweetheart had started gathering up children and hats and the group moved slowly back toward Loflin's truck, each of us shredding first trails through the cannabis farm. When we reached a spot beyond the hemp field's boundaries, Loflin pointed to a patch of altogether more desert-like dirt than what we'd been mucking through in the field. "This year's crop is already clearly stabilizing the soil, which is another thing I'm trying to show my neighbors. That and the good news about the water."

"What's the good news about the water?"

"The hemp's wanting about a third to half the water of corn, which is the dominant crop here now," he said. "Twelve to fourteen inches for hemp, versus twenty-four to thirty-six for corn." This was big news for farmers watering from the declining Ogallala Aquifer underneath us. "Everyone knows how much their well can produce—folks could dryland-crop hemp."

Dusk was coming on fast. "Any other advice for farmers moving to hemp next season and in coming years?" I asked.

Loflin stopped at his truck door. "Farming is farming," he said, sounding exactly like Grant Dyck, the Manitoban hemp farmer.

"Every new crop is definitely a learning curve. Realize that, do your research, and you might have a little fun along the way."

Yes, this makes about half a dozen sources in this book who have included "have fun" in their entry-level business advice. Still, driving back from the field, I remembered Dyck's combine fires. "Are you ready for harvest?" I asked.

Loflin rolled his eyes and fidgeted uncomfortably in the driver's seat. "I'm consulting with Anndrea Hermann," he muttered. "I know with the oil crop we have to consider moisture contact at harvest, plus aeration during storage. Just more parts of the process to learn. Anyway, Anndrea's coming down here at harvesttime."

"Smart move," I said, speaking as a hemp journalist.

"We were just talking today about whether to turn off his irrigation now or later," Hermann told me when I called her to ask for an assessment of Loflin's effort. "I told him I thought he could keep watering for a bit. And I saw samples of the seeds—they look like they're forming well. I think he'll do fine."

One detail that Hermann found interesting, from observing one of Colorado's first-ever hemp fields: "The crop's growing like we see on the high plains of Alberta." That might help farmers choose cultivars to use or hybridize in future years.

Add Loflin to the list of Americans grateful for Canada's hemp know-how. After all, how could he know what to do? Forget about being Colorado's first hemp field. Almost no one has cultivated the plant south of the forty-fifth parallel in more than half a century. Billion-dollar industries have to start somewhere.

And Loflin did believe he was in on the ground floor of such an industry. In fact, he was happy to get off the topic of harvest, which appeared like it might be causing him to lose a little sleep. "Two years from now we'll already be talking about the development of a major commercial industry that's well under way, from these seeds," Loflin predicted. "Between the construction industry, seed oil, and building America's seed bank, Colorado is going to be leading the hemp revolution."

While my sleepy kids were wiggling into their own car seats back at the Loflin farmhouse, Ryan disclosed that he was "a little apprehensive" about the lingering federal law quagmire. "I don't think they'll bother us. Not with drug cartels coming up from Mexico. If they do they'll look pretty foolish."

If the DEA's evolving public statements were any indication, he was probably right. Notice the considerably less bellicose tone in the second of these two quotes from the same agency, in major media six months apart.

It really doesn't matter whether it looks different or it looks the same. If it's the cannabis plant, it's in the Controlled Substances Act and, therefore, enforceable under federal drug law. —DEA special agent Paul Roach, threatening Michael Bowman on NPR's *Morning Edition*, January 28, 2013

Hemp farmers are not on our radar. —Denver DEA spokesman, in a *New York Times* article about Loflin, August 5, 2013

This essentially reflects America's widespread and growing support for hemp. Pulling a leaf from his own field off the skier on his shirt, the last thing this technical federal felon told me was, "I want to build an industry for America—something my kids and their kids can rely on." Then he gave me a gallon jug of Dr. Bronner's Magic Soap (peppermint) as a parting gift. As I loaded it into my rig, I thought, *Next bottle: American-grown.*

Periodic bolts of violet lightning on the southern horizon reminded me that this year at least promised to be a wet one—an auspicious rainy season for the inaugural modern hemp crop, from which it's possible much of the early American commercial seed market will descend.

Just as I was starting my engine, Loflin waved to a neighbor driving by in front of the property. He knew everyone in Springfield. This of course matters in a small town, and reminded me to investigate whether Springfield, Colorado shared Byers, Colorado's readiness to grow hemp en masse. Which is to say, in large enough quantities to support a local processor.

So instead of pulling out of Springfield, I backtracked and stopped in at Pappy's BBQ (get the fried okra) in the part of town that still had its original 1886 edifices. There, on a Tuesday evening, I was able to confirm Loflin's sense that the town was on board the hemp train.

"I don't got a problem with it," said lifelong resident Jack Carson, sixty-one, from a booth he shared with his wife, Debra. Along the wall above them was a shelf filled with photos of Baca County veterans who had served in long-ago and current conflicts. "It's a cash crop for farmers, and Lord knows we need it."

"I know the Loflins," Debra chimed in. "Ryan went to school with my daughter. They're good people."

Another family expressed the same view while I waited for my own clan's order. It stuck me viscerally that hemp fit right in in Baca County, Colorado. It fits right in in the heartland. *God bless America*, I thought, aiming the rig southwest and home.

ACKNOWLEDGMENTS

*A*s I wistfully finish this project, I'm sending out heartfelt thanks to all the hemp experts and consultants who gave freely of their time and info during its research: Anndrea Hermann, Bill Althouse, and Michael Bowman, in particular, went beyond the call. Also incredibly helpful were Ian Pritchett, Adam Eidinger, Michael Carus, John Hobson, Tim Callahan, Greg Flavall, Kelly and Bob King, David Bronner, Grant and Colleen Dyck, Agua Das, Simon Potter, Farhoud Delijani, Ryan Loflin, Don Wirtshafter, Steve Levine, Jillane Hixson, Dave Tzilkoski, Adrian Clark, Norm Roulet, Ellen Komp, Shaun Crew, Danielle Schumacher, and, well, everyone you just met in this book.

My biggest thanks, as always, goes to my superlatively supportive family, who milked the goats while I was jamming on deadline (endless hug payback already on its way). Gracias also to Mike Behar for seeing the potential of this project after discovering his own toddler reading a review of my previous book on the potty, and to Markus Hoffmann, for his usual kindness and professionalism. And thanks to Leigh Huffine and the Chelsea Green team for help with the live events. My editor, Brianne Goodspeed, was terrific to work with. Also, I'm beaming intercontinental appreciation to Michel Degens and Derrick Bergman, who set up several of my European site visits.

And, finally, thanks to the initial fellow or lady who, probably after watching the local animals munch it, first snapped off a branch of *Cannabis sativa* and thought, *Ya know, this'd make a great roof/ sandal/basket/food/rope/medicine/party gift*—and then two weeks later had a follow-up thought: *This is, like, the King of Plants, in terms of usefulness. Hey, pass the mastodon burgers.*

NOTES

1. This after Colorado voters amended their constitution to allow all forms of cannabis on November 6, 2012. Hemp was explicitly specified in that Amendment 64.
2. For fans of obscure government regulation, there's some indication that this might have already happened for non-edible (fiber, rather than seed) versions of hemp, back in 2003. According to Kentucky attorney Luke Morgan (a white-shoe Bluegrass State lawyer, he used to work in the state attorney general's office), quoted in the Kentucky publication *The Lane Report* on August 6, 2013, Drug Enforcement Agency Final Rule (FR Doc 03-6805) "Exemption From Control of Certain Industrial Products and Materials Derived From the Cannabis Plant" frees cannabis-plant-derived industrial products and feeds not intended for human consumption from federal control under the Controlled Substances Act.
3. "Hemp: A Confusing, Historical, and Fascinating Plant," Canadian Hemp Trade Alliance, accessed October 18, 2013, www.hemptrade.ca/index.php.
4. West told me he learned this tale from James F. Hopkins's *A History of the Hemp Industry in Kentucky*, a 1951 book reissued by the University of Kentucky Press in 1998.
5. In fact, as I send this book to my publisher, a source close to the FARRM Bill negotiations has emailed me to say I will come off like a prophet if I predict that the final hemp amendment wording therein will wind up allowing even stronger cultivation allowances than the "university research" wording that passed the House in 2013. Key Republicans and Democrats have agreed to present it as a states-rights issue, my source told me. "We're talking about full legalization," he said. Well, prediction is free, so, OK, I hereby predict that.
6. Hemp as an industrial-scale tree-free paper option today is promising enough that International Paper Corporation has looked into it. The good folks at my publisher, Chelsea Green Publishing of White River Junction, Vermont, perhaps in response to my nonstop ranting on the subject, pulled out all the stops—including pursuing sources in Canada and China until the eleventh hour—in trying to print the first edition of *Hemp Bound* on hemp paper. It was, my editor reported to me with deeply furrowed brow, simply not cost feasible (YET!) on a mass scale. Stay tuned.

7. President Obama already wore a Colorado-made hemp scarf on a campaign stop in 2012. He knows the deal.

8. The cannabis plant actually has three main varieties: *Cannabis sativa, C. indica,* and *C. ruderalis.* In Canada, the government says you can cultivate any of these industrially, provided that "the leaves and flowering heads . . . do not contain more than 0.3% THC" (Industrial Hemp Regulations, SOR/98-156, Controlled Drugs and Substances Act, Government of Canada, last modified October 1, 2013, accessed October 18, 2013, http://laws-lois.justice.gc.ca/eng/regulations/SOR-98-156/FullText .html). THC: That's the substance that allowed Woodstock, and that *not more than 0.3%* of it will be our definition of hemp and industrial cannabis in this book.

9. Some varieties of hemp might also possess many of the medicinal and health maintenance properties that psychoactive cannabis has, only without the psychoactive elements. This might very well prove a lucrative market for industrial cannabis, but one into which, having discussed it at length in *Too High to Fail*, my previous book, I won't delve in these pages.

10. Ernest Small and David Marcus, "Hemp: A New Crop with New Uses for North America," in *Trends in New Crops and New Uses*, eds. Jules Janick and Anna Whipkey (Alexandria, VA: ASHS Press, 2002).

11. Beverly Fortune, "Advocates of Industrial Hemp Point to Kentucky's Past as Top Producer," *Lexington Herald-Leader*, January 1, 2013.

12. Governor Brown finally signed hemp legalization into California law on September 25, 2013.

13. Renée Johnson, "Hemp as an Agricultural Commodity" (working paper, Congressional Research Service, Washington, DC, July 24, 2013).

14. Agua Das and Thomas B. Reed, "United States: Biomass Fuels from Hemp—Seven Ways Around the Gas Pump," *restore's blog* (blog), *Hemp News*, October 14, 2011, www.hemp.org/news/content/biomass-fuels-hemp.

15. This is actually the easiest time in history to cover U.S./Canadian business issues, because in 2013 as I write these words the two nations' dollars are worth about the same.

16. Erin M. Goldberg, Naveen Gakhar, Donna Ryland, Michel Aliani, Robert A. Gibson, and James D. House, "Fatty Acid Profile and Sensory Characteristics of Table Eggs from Laying Hens Fed Hempseed and Hempseed Oil," *Journal of Food Science* 77 (2012): S153–60. doi: 10.1111/j.1750-3841.2012.02626.x.

17. Of which sixteen million dollars went to farmers in 2012.
18. "List of Approved Cultivars for the 2012 Growing Season," Health Canada, last modified October 15, 2012, accessed October 21, 2013, www.hc-sc.gc.ca/hc-ps/pubs /precurs/list_cultivars-liste2012/index-eng.php.
19. www.hemptechnology.co.uk/agronomy.htm.
20. Agua Das and Thomas B. Reed, "United States: Biomass Fuels from Hemp— Seven Ways Around the Gas Pump," *restore's blog* (blog), *Hemp News*, October 14, 2011, www.hemp.org/news/content/biomass-fuels-hemp.
21. Though this is projected to change in coming seasons, today Canadian hemp farmers overwhelmingly grow for seed oil only. Some bale their hemp straw and sell it locally for animal bedding.
22. "Hemp's Future in Chinese Fabrics," International Year of Natural Fibres, Food and Agriculture Organization of the United Nations, accessed October 18, 2013, www.naturalfibres2009.org/en/stories/hemp.html.
23. This is a yield the Canadian Hemp Trade Alliance trade group says is rising as farmers like Dyck get out those kinks.
24. Technically, industrial cannabis isn't illegal: The DEA can issue hemp cultivation permits. Exemptions from the Controlled Substances Act, essentially. Good luck. It's happened a handful of times during the drug war, including for West's research plot in Hawaii. Most efforts have been stifled. Alex White Plume, believing he lived on sovereign territory not subject to U.S. federal law, tried to grow a hemp crop on the Oglala Lakota Nation (Pine Ridge Reservation, South Dakota). After being raided by armed DEA agents who destroyed the crop in 2000 and 2001, in 2002 Plume got a harvest to market, though on the morning of harvest he was charged with eight federal civil violations, according to the documentary *Standing Silent Nation*. The North Dakota saga Goehring referred to involved the effort by state Republican lawmakers, farmers, and the state's then agricultural commissioner to acquire DEA hemp cultivation permits in 2007. They were ignored by our federal public servants for three years, even after they sued for an answer. A federal court finally ruled that they'd have to be raided before they could challenge the DEA. They decided to hold off planting.
25. Lime Technology has a proprietary mixture spelled "Hemcrete." We'll be using the generic "hempcrete." You can, after all, mix hemp hurds and lime yourself if you want, or even form a company that produces it commercially.

26. It lives under the "documents" link at eiha.org.

27. Lynn Osburn and Judy Osburn, "Hemp Plywood Becomes a Reality," *North Coast Xpress*, February–March 1999, 9.

28. "World Crude Oil Consumption by Year," Index Mundi (website), accessed October 22, 2013, www.indexmundi.com/energy.aspx.

29. I can attest to this, having driven my own truck on the local Chinese food joint's waste oil for seven years.

30. Russ Bellville, "Could Hemp Help Nuclear Clean-up in Japan?" *Examiner* (blog), March 13, 2011, www.examiner.com/article/could-hemp-help-nuclear-clean -up-japan.

31. Beverly Fortune, "Advocates of Industrial Hemp Point to Kentucky's Past as Top Producer," *Lexington Herald-Leader*, January 1, 2013.

32. Associated Press, "Feldheim, German Village, Powered by Renewable Energy," *Huffington Post*, December 29, 2011, www.huffingtonpost.com/2011/12/29/feldheim -germany-renewable_n_1173992.html.

33. The plant wall's polymer, different from edible lignan.

34. The first Levis dungarees were made of hemp.

35. A 2013 University of Kentucky study somewhat concurred, further pointing out that even high seed oil prices could evaporate when more producers come online. The study still predicted revenue for Kentucky hemp farmers of more than $300 per acre under certain sets of economic conditions, while maintaining those earn-ings from conventional corn/soy farming would likely stay higher, at least initially. (Lynn Robbins, Will Snell, Greg Halich, Leigh Maynard, Carl Dillon, and David Spalding, "Economic Considerations for Growing Industrial Hemp: Implications for Kentucky's Farmers and Agricultural Economy" [working paper, Department of Agricultural Economics, University of Kentucky, July, 2013], https://www .google.be/search?q=Economic+considerations+for+growing+industrial +hemp&ie=utf-8&oe=utf-8&rls=org.mozilla:en-US:official&client=firefox-a &gws_rd=cr&ei=1vGIUp7-BoTQtQbQ9YCgCw.)

36. "Hemp & Lime Construction," Hemp Technology, accessed October 18, 2013, www.hemptechnology.co.uk/hemcrete.htm.

37. Hossein Shapouri and Michael Salassi, "The Economic Feasibility of Ethanol Production from Sugar in the United States" (working paper, U.S. Department of Agriculture, Washington, DC, July 2006).

38. *History of the Ohio Falls Cities and Their Counties* (Cleveland, 1882).

39. Vote Hemp's Eric Steenstra actually has an answer to that question: "Change at the state level has been crucial," he said. "If your state has passed hemp cultivation legislation, push your state officials toward implementing it. If you live in the forty states that don't yet have legal hemp farming, push your state officials toward passing it." He also suggested calling your federal representatives to support two congressional bills (S359 and HR525) that will fully allow hemp cultivation, which as we've discussed is stronger than the "university research" wording passed by the U.S. House of Representatives in 2013. "And buying hemp products is very important," Steenstra told me. "Economic development is going to move this issue politically."

40. Zachary Barr, "Hemp Gets the Green Light in New Colorado Pot Measure," *Around the Nation* (blog), National Public Radio website, January 28, 2013, www.npr.org/2013/01/28/170300215/hemp-gets-the-green-light-in-new-colorado-pot-measure.

41. At one point during my Canadian research, a local RCMP officer called Hermann to tell her a letter of approval she was seeking for another farmer's hemp crop was ready. "Law enforcement is here to help us," she said.

42. Tashajara Stenvall, Julian Stickley, Bryant Nagelson, Piauwasdy, Izy Mcclure, and Sarah Meanwell, "Cannabis: The Ethnobotany and Political Ecology of Hemp," *Environmental and Food Justice* (blog), December 28, 2012, http://ejfood.blogspot.com/2012/12/ethnoecology-blogs-autumn-2012-cannabis.html.

43. As described in my *Too High to Fail: Cannabis and the New Green Economic Revolution*.

44. Ernest Small and David Marcus, "Hemp: A New Crop with New Uses for North America," in *Trends in New Crops and New Uses*, eds. Jules Janick and Anna Whipkey (Alexandria, VA: ASHS Press, 2002).

45. Steve Raabe, "Hemp Industry Poised to Grow in Colorado with New Legal Status," *Denver Post*, January 14, 2013.

46. May told me he was willing and in fact planning to plant three acres in 2013, but the seed, which would have come from Bowman, didn't arrive when Bowman delayed his own farm's planting a year.

47. On a visit to a Belgian hemp farm, Ingrid Maris, a high school teacher, told me she had begun cultivating on her family's land in economically struggling

Limburg province for the same reason as Loflin: "I want to demonstrate to my neighbors, who are traditionally minded farmers, that hemp is viable, has so many applications, and, of course, is better for the soil than monoculture," she said as we toured the idyllic Flanders countryside. "It grows very fast on its own with no pesticides and chemical additives."

48. He's dismantled and resold a hundred barns.

49. Alan Haney and Benjamin B. Kutscheid, "An Ecological Study of Naturalized Hemp (*Cannabis savita* L.) in East-Central Illinois," *American Midland Naturalist* 93(1) (University of Notre Dame, January 1975).

50. Among the roughly two dozen Colorado farmers who planted industrial cannabis in 2013, none was as open as Loflin and Bowman. The reasons for this go beyond the inclination to wait for the implemention of state cultivation regulations for the 2014 season. Chris Boucher, a founder of the Hemp Industry Association and a hempster since the 1990s, said the paced start results from a market that needs a few years to develop. At a company called U.S. Hemp Oil, where he's vice president of product development, the plans are to "ramp up production in Colorado in baby steps, to see what works at seven thousand feet on the thirty-ninth parallel." As many as two hundred farmers will be planting in 2014, and the number could easily surpass one thousand by 2016 if enough seed is available, according to Bowman.

RESOURCES

The list of players in all of these categories is going to be growing exponentially in coming months and years. These resources are intended just to get the interested hemp farmer or entrepreneur started. It's written at the tail end of the era when the professional hemp world was a small family.

HEMP POLITICAL ADVOCACY AND INDUSTRY GROUPS
The following organizations can provide resources useful in working toward hemp cultivation and processing in your region:

> Vote Hemp: www.votehemp.com/write_congress.html
> Hemp Industries Association (HIA): www.thehia.org
> Canadian Hemp Trade Alliance: www.hemptrade.ca
> European Industrial Hemp Association: www.eiha.org
> Kentucky Hemp Growers Cooperative Association: www.kentuckyhempgca.org
> Rocky Mountain Hemp Association: http://rockymountainhempassociation.org
> BioFibre Conference, Manitoba: http://biofibe.com
> Hemp History Week: www.hemphistoryweek.com

HEMP SEED
> Ryan Loflin's Rocky Mountain Hemp: http://rockymountainhempinc.com
> Ben Holms's Centennial Seed Distributors: www.centennialseeds.com
> The Canadian government's official hemp cultivar list: www.hc-sc.gc.ca/hc-ps
> /pubs/precurs/list_cultivars-liste2012/index-eng.php

UNIVERSITY COURSE
> Oregon State University Hemp Course, Corvallis, Oregon: http://bit.ly/XsChVe

HEMP SEED OIL
> Manitoba Harvest: manitobaharvest.com
> Hemp Oil Canada (processing for wholesalers): www.hempoilcan.com
> Nutiva: nutiva.com

Hemp Building

American Lime Technology: www.americanlimetechnology.com
A hemp building conference: http://internationalhempbuilding.org
/events/4th-international-hemp-building-symposium
Greg Flavall and Hemp Technologies' North Carolina house:
www.youtube.com/watch?v=eZbYsMsMW4Q
The house Tim Callahan designed: http://alembicstudio.com/portfolio
/the-nauhaus-prototype

Hemp Textiles

Fabrics, webbing, rope, et cetera:
www.envirotextile.com
www.hemptraders.com
www.pickhemp.com

Retail fabrics, webbing, trim, et cetera:
www.nearseanaturals.com

Retail finished goods:
http://rawganique.com
www.etsy.com/shop/gaiaconceptions
http://uprising.be/

Hemp for Energy Independence

In the end, energy independence is going to come via energized localities.

A bona fide small-farm-sized biomass gasification engine at work in Austria:
www.youtube.com/watch?v=0P7zFw_xff0
A California biomass plant in action (with wood chips for now):
http://ucanr.edu/blogs/blogcore/postdetail.cfm?postnum=7885

Other Hemp Industries

Hemp Plastic Bottle Kickstarter Project: http://hempwaterbottles.tripod.com
Hemp Shield eco-friendly wood sealer: www.hempshield.net
Hemp Decorticator: www.youtube.com/watch?v=U8LFErsq6wl

Hemp cereal at the International Space Station: http://holycrap.ca/whats-new
/holy-crap-cereal-rockets-to-the-international-space-station

U.S. train running on vegetable oil today—hemp tomorrow? http://illianaroad
.com/railroading/railway-powered-by-waste-vegetable-oil

Lignol, a cellulosic industrial process for which hemp might prove ideal:
www.lignol.ca

A Dutch company providing hemp for everything from home construction to
animal-care products to BMW door panels (company motto: Nature Wins!):
www.hempflax.com/en

The Canadian Composites Innovation Centre's FibreCITY franchising system:
http://fibrecity.ca/about.html

Tour Boston's Hemp History: http://hempology.org

A website with links to manifold hemp industry sectors:
www.hemp-technologies.com

A Czech company growing hemp and making it into hemp snacks (company
motto: Love stemming from respect for the traditions of our ancestors, while
also taking care of our children): www.hempoint.cz

A Beverly Hills company selling hemp rugs: www.caravanrug.com
/hemp-rugs-cat-299

HEMP CONSULTANTS

The putative hemp entrepreneur on both the production end and the finished-prod-
uct side will have met several consultants in these pages. There are others. As every
hemp farmer I spoke with urges, do your research before leaping into any business.
When it comes to outside help for your endeavor, the same applies.

Anndrea Hermann, The Ridge International Cannabis Consulting:
www.facebook.com/pages/The-Ridge-International-Cannabis-Consulting
-Anndrea-Hermann-204-377-4417/219657108125220

Agua Das, longtime hemp researcher: das.ellis@gmail.com;
www.hemphasis.net/Notable/notable_files/das.htm

Jason Lauve for Professional Industrial Hemp Consultation: www.lauve.com

Michael Carus, European Industrial Hemp Association:
michael.carus@eiha.org; www.eiha.org

John Hobson, Hemp Technology (UK):
　　john@hemptechnology.co.uk; www.limetechnology.co.uk
Erik Hunter, Rocky Mountain Hemp Association:
　　erik@rockymountainhempassociation.org;
　　http://rockymountainhempassociation.org

RECENT HEMP DOCUMENTARIES

Government Grown: http://governmentgrownhemp.weebly.com
Bringing It Home: www.bringingithomemovie.com
Hemp researcher David West on feral hemp in Nebraska:
　　www.youtube.com/watch?v=R5IQ2PrSWG4

HISTORICAL HEMP DOCUMENTARY

Hemp for Victory, the 1942 U.S. government pro-hemp film:
　　www.youtube.com/watch?v=a1oFcgLfgV0

HEMP BOOK

Must read for the history of industrial cannabis: *Cannabis*, Mathias Broeckers,
　　The Hash Marihuana Hemp Museum Press, Amsterdam, 2002

INDEX

FSC
www.fsc.org

MIX

Paper from
responsible sources
FSC® C013483

Chelsea Green Publishing is committed to preserving
ancient forests and natural resources. We elected to print
this title on 100-percent postconsumer recycled paper,
processed chlorine-free. As a result, for this printing, we
have saved:

52 Trees (40' tall and 6-8" diameter)
23 Million BTUs of Total Energy
4,465 Pounds of Greenhouse Gases
24,215 Gallons of Wastewater
1,621 Pounds of Solid Waste

Chelsea Green Publishing made this paper choice because
we and our printer, Thomson-Shore, Inc., are members
of the Green Press Initiative, a nonprofit program dedicated to supporting authors, publishers, and suppliers
in their efforts to reduce their use of fiber obtained
from endangered forests. For more information, visit:
www.greenpressinitiative.org.

Environmental impact estimates were made using the Environmental Defense Paper Calculator.
For more information visit: www.papercalculator.org.

ABOUT THE AUTHOR

*D*oug Fine is a comedic investigative journalist, bestselling author, and solar-powered goat herder. Since emigrating from suburbs to wilderness in the 1990s, he has reported from five continents for the *Washington Post*, *Wired*, *Salon*, the *New York Times*, *Outside*, National Public Radio, and *U.S. News & World Report*. His work from Burma was read into the *Congressional Record* (by none other than pro-hemp senator Mitch McConnell), and he won more than a dozen Alaskan press club awards for his radio reporting from the Last Frontier. Fine is the author of three previous books: *Too High to Fail: Cannabis and the New Green Economic Revolution*; *Farewell, My Subaru: An Epic Adventure in Local Living*; and *Not Really an Alaskan Mountain Man*. A website of his print work, radio work, and short films is at www.dougfine.com. Twitter: organiccowboy.

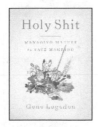

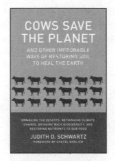